THIS IS PHILOSOPHY OF SCIENCE

THIS IS PHILOSOPHY
Series editor: Steven D. Hales

Reading philosophy can be like trying to ride a bucking bronco – you hold on for dear life while "transcendental deduction" twists you to one side, "causa sui" throws you to the other, and a 300-word, 300-year-old sentence comes down on you like an iron-shod hoof the size of a dinner plate. *This Is Philosophy* is the riding academy that solves these problems. Each book in the series is written by an expert who knows how to gently guide students into the subject regardless of the reader's ability or previous level of knowledge. Their reader-friendly prose is designed to help students find their way into the fascinating, challenging ideas that compose philosophy without simply sticking the hapless novice on the back of the bronco, as so many texts do. All the books in the series provide ample pedagogical aids, including links to free online primary sources. When students are ready to take the next step in their philosophical education, *This Is Philosophy* is right there with them to help them along the way.

This Is Philosophy, Second Edition
Steven D. Hales

This Is Philosophy of Mind
Pete Mandik

This Is Ethics
Jussi Suikkanen

This Is Political Philosophy
Alex Tuckness and Clark Wolf

This Is Business Ethics
Tobey Scharding

This Is Metaphysics
Kris McDaniel

This Is Bioethics
Ruth F. Chadwick and Udo Schuklenk

This Is Philosophy of Religion
Neil Manson

This Is Epistemology
J. Adam Carter and Clayton Littlejohn

This Is Philosophy of Science
Franz-Peter Griesmaier and Jeffrey A. Lockwood

Forthcoming:

This Is Environmental Ethics
Wendy Lee

This Is Philosophy of Mind, Second Edition
Pete Mandik

This Is Modern Philosophy
Kurt Smith

THIS IS

PHILOSOPHY

OF SCIENCE

AN INTRODUCTION

FRANZ-PETER GRIESMAIER

JEFFREY A. LOCKWOOD

WILEY Blackwell

The right of Franz-Peter Griesmaier and Jeffrey A. Lockwood to be identified as the authors of this work has been asserted in accordance with law.

Registered Office
John Wiley & Sons, Inc., 111 River Street, Hoboken, NJ 07030, USA

Editorial Office
111 River Street, Hoboken, NJ 07030, USA

For details of our global editorial offices, customer services, and more information about Wiley products visit us at www.wiley.com.

Wiley also publishes its books in a variety of electronic formats and by print-on-demand. Some content that appears in standard print versions of this book may not be available in other formats.

Library of Congress Cataloging-in-Publication Data
Names: Griesmaier, Franz-Peter, author. | Lockwood, Jeffrey Alan, 1960- author.
Title: This is philosophy of science : an introduction / Franz-Peter Griesmaier, Jeffrey A. Lockwood.
Description: First Edition. | Hoboken, NJ : John Wiley & Sons, 2022. | Includes bibliographical references and index.
Identifiers: LCCN 2021044855 (print) | LCCN 2021044856 (ebook) | ISBN 9781119757993 (Paperback) |
  ISBN 9781119758013 (PDF) | ISBN 9781119758006 (ePub)
Subjects: LCSH: Science--Philosophy.
Classification: LCC Q175 .G745 2022 (print) | LCC Q175 (ebook) | DDC 501--dc23/eng/20211119
LC record available at https://lccn.loc.gov/2021044855
LC ebook record available at https://lccn.loc.gov/2021044856

Cover Image and Design: Wiley

Set in 10/12pt and Minion Pro by Integra Software Services Pvt. Ltd, Pondicherry, India

SKY10033436_022822

CONTENTS

PREFACE

More in need of philosophy are the sciences where perplexities are greater.
— Carlo Rovelli, Codeveloper of loop quantum gravity

You are an unusual student. We' d be surprised if 1% of STEM majors (or even most philosophy majors) take a course in the philosophy of science. And this is an unusual book in three ways which we hope will provide a readable college text.

First, most philosophy of science texts are written for philosophy majors. While this book will provide them with a solid foundation, our goal is to provide STEM majors with a relevant book. But why would they seek such a text?

Imagine that there is an opera singer who hits every note, but she has no understanding of the composition. Is she truly a musician or just a performer? In German, a Musiker has a deep understanding of music, while a Musikant merely plays music. Understanding the philosophy of science is necessary to become a scientifiker rather than a scientifikant (these terms don't really exist but you get the idea).

Second, the coauthors are an odd couple. FPG is an immigrant and first-generation college student whose educational path led him to teach philosophy to hundreds of students. JAL was a faculty member in entomology for nearly 20 years before joining the philosophy department. As such, the authors have been there, with "there" being a classroom where you wonder if you belong and an academic setting where you are an outsider.

This text would've been easier to write alone, without the philosopher nitpicking the scientist's versions of concepts or the scientist revising the philosopher's accounts of empirical research. We crafted this book together so as to respect the intellect of the student and recognize the unfamiliar practice of thinking about the nature of knowledge and reality.

Third, this book is a kind of chimera structured around two of philosophy's core disciplines – epistemology and metaphysics (an unconventional framework) – with lots of historical and contemporary examples from the sciences, including the physical (from cosmology to geology), biological (from ecology to medicine), and social (from anthropology to economics). We recognize that the science–philosophy interface requires intriguing case studies (e.g., time travel) to illustrate abstract concepts because if a student is not engaged, then the book is just filling space.

We should also note what is missing from this text. Apart from the need to forgo covering every important view on the topics we chose to treat (such as the kairetic account of explanation), there is also no systematic treatment of ethical, social, religious, and political questions. This absence is not intended to suggest that the cultural context of science is unimportant, but that an exploration of the values–science interface warrants an entire text.

We agree with Carlo Rovelli, theoretical physicist writing for Scientific American, who argued that:

> Philosophy provides guidance how research must be done. Not because philosophy can offer a final word about the right methodology of science [but] because the scientists who deny the role of philosophy in the advancement of science … are the ones trapped in the ideology of their time.

But not only science should take account of philosophy. As Rovelli continues: "Just as the best science listens keenly to philosophy, so the best philosophy listens keenly to science." We hope that this text fosters such mutual respect among students majoring in philosophy and STEM fields.

ACKNOWLEDGMENTS

We thank our department head, colleagues, and families for their encouragement and support in the writing of this book. We especially want to thank the students in FPG's philosophy of science courses at the University of Wyoming, who provided valuable feedback on earlier versions of the manuscript that eventually morphed into this book. Marissa Koors from Wiley suggested the inclusion of our manuscript in their series This is Philosophy, and the series editor, Steven Hales, got the project going. Our thanks to both of them, as well as to Charlie Hamlyn, who saw the book through to completion. Five anonymous reviewers helped us improve the manuscript by alerting us to several errors and by suggesting discussions of additional topics to enlarge the scope of the text, for which we are grateful. A number of former students and colleagues contributed directly to the content, either by critically evaluating some parts of the manuscript, or by helping us see some of the issues in a different light during fruitful discussions. In particular, our thanks go to Elizabeth Cantalamessa, Conor O'Malley, Bradley Rettler, and Lindsay Rettler. Thanks also to Elizabeth Bell, whose delightful illustrations will help the reader get a better grasp of some of the more abstract topics, and to John Poland for compiling the index.

ABOUT THE COMPANION WEBSITE

A website for *This is Philosophy of Science* can be accessed at https://thisisphilosoph.wordpress.com. But why would you want to go to a website when you already have a textbook and you have more than enough reading and studying to fill your days? Well, we suggest that the website is a valuable resource when:

You've been given an assignment–perhaps a paper to write or presentation to give–and you sure don't want to mess up technical terms. The website Glossary is an ideal resource to assure that when you write "emergent properties" or say "empiricism" you don't rely on a philosophically flawed use of the words.

The exam is just days away. How can you best prepare? First, go to the website Glossary and make sure you've fully mastered the basic concepts that appear in the chapters covered by the test. And second, test yourself using the multiple-choice questions that appear on the website (these represent many of the important topics covered in each chapter).

As the instructor of the course, tapping into the chapter quizzes on the website might provide an ideal basis for organizing a study session for your students–and perhaps even telling them that your exam will include questions drawn directly from the website!

1

PILLARS OF SCIENCE: REASONS, KNOWLEDGE, AND TRUTH

The empirical sciences are dedicated to describing, explaining, and predicting natural phenomena, such as ionic bonding, predation rates, formation of galaxies, bird migrations, gravitational waves, protein metabolism, or the probability of an earthquake in northern California. The ultimate goal is to gain *genuine knowledge* about such phenomena – what exactly they are, what regularities govern their occurrence, why and how they occur, when they might happen in the future, and with what probability.

In subsequent chapters, we will investigate many of the difficult and fascinating issues involved in a successful exploration of the natural world. Some of those issues arise with equal force in different sciences, while others may arise in only one discipline, but not in any of the others, at least not in the same form. For now, however, we are going to discuss some of the concepts and procedures that arguably can be found in all empirical sciences. These include (i) the fundamental nature of reasons we give for accepting or rejecting a scientific hypothesis, or any ordinary claim about the world for that matter; (ii) the general structure of the different kinds of inferences we can make on the basis of observations; (iii) the nature of truth and the specter of relativism; and (iv) the relation between facts, hypotheses, laws, and theories. Having a good grasp of these concepts will prove invaluable for understanding the intricacies of empirical science and the philosophical problems that arise in the course of scientific investigations.

The initial discussion will be a bit on the abstract side, and you might at first find it puzzling how questions such as the ones we are going to raise in

This is Philosophy of Science: An Introduction, First Edition. Franz-Peter Griesmaier and Jeffrey A. Lockwood.

this chapter are of any relevance to the practice of science. The situation is perhaps similar to learning the rules of baseball before swinging a bat. Sure, you might hit the ball if you don't know the how the game is played, but you might well run in the wrong direction afterward! And so, as our discussion proceeds in the later chapters, you will find that having gained mastery of some fundamental concepts in epistemology (the study of knowledge) is of great value. So let's get going.

1.1 Epistemic Reasons

Typically, when we wonder whether we should accept some claim, such as a scientific hypothesis, or not, we look for reasons for doing so. Another way of putting this is that we don't believe something without having reasons. For example, if someone asks you to accept that there are intelligent, extraterrestrial life forms, you're likely to ask for reasons before you adopt this belief. And if the other person just hems and haws, you're not going to believe in extraterrestrials. On the other hand, if you are presented with the cosmic background radiation as evidence for the occurrence of the so-called Big Bang, you have some reason for believing that the universe emerged through this sort of process. (Notice that when we use the term "belief," we do not mean to talk about religious faith.)

Of course, not any old reason is a good reason. If I believe something because I would like it if it were true, or because I am better off believing it, the belief might well turn out to be false – and in the overwhelming number of cases, it will be. Wishful thinking rarely leads to true beliefs. Thus, we need another kind of reason for believing something if we want to find out the truth about the world.

Reasons of the desired kind are called *epistemic* reasons. They are the sort of reasons that allow us to accept a belief only if there is good evidence for its truth, or only if the belief doesn't contradict other, already well-established beliefs derived from good evidence. Of course, it is very contentious what makes for a *good* epistemic reason. The debate over which (types of) epistemic reasons are to be preferred over others constitutes part of what's called *epistemology*, or the study of knowledge. Epistemic reasons are usually divided into two kinds: those which *guarantee*, in a sense to be specified momentarily, the truth of their target beliefs, and those that merely make the truth *more probable*. We start with the former.

1.1.1 Conclusive Reasons

The first type of epistemic reasons are called *conclusive* reasons. A reason (R) is conclusive for some belief (B) if and only if the belief *B must be true if R is true*. And this condition holds even if there is not just one reason for B, but also in cases in which B rests on many reasons. In more general terms, if all the reasons for a belief are true, and if they are conclusive reasons, then their target belief *must* be true. Conclusive reasons guarantee true beliefs, which is strongest basis one can have for believing something. So, how can we understand this definition?

A good example for conclusive reasons are the premises of a *deductively valid argument*. In such an argument, if all the premises are true, then the conclusion *must* be true as well. Here's a simple example:

Premise 1:   All humans are mortal.
Premise 2:   Stephen Hawking is human.
Conclusion: Thus, Stephen Hawking is mortal.

Clearly, if premises 1 and 2 are both true, then the conclusion is guaranteed to be true. Thus, the two premises together are conclusive reasons for believing that Hawking is mortal. But this sort of reasoning is not often helpful for advancing our scientific understanding of the world. Let's see why.

Notice that in a deductive argument, what's really happening is that information that is already contained implicitly in the premises, is made explicit in the conclusion. In other words, the conclusion does not reveal any new information. It restates the information that's already contained in the premises. That's why such inferences are *safe*: truth in – truth out.

Deductive reasoning (i.e., reasoning that proceeds by providing conclusive reasons) is mostly confined to two major disciplines: mathematics and logic. Yes, sometimes we use deductive reasoning in the empirical sciences, such as in cases in which we deduce observational consequences from a theory in order to test it (which can include refuting the theory):

Premise 1 (Theory 1)	All birds can fly.
Premise 2 (Theory 2)	Penguins are birds.
Premise 3 (Deduced Consequence)	Penguins can fly.
Premise 4 (Observation):	Penguins can't fly.
Conclusion:	Not all birds can fly.

However, a lot of scientific reasoning is nondeductive. Why? Because typically, in scientific reasoning, we want to infer something about the world at large on the basis of a limited number of observations. Such inferences are inherently risky because their conclusions convey information that goes beyond the information contained in the descriptions of the actual, limited observations that have been made.

For example, if I infer, on the basis of having observed the eating habits of 20 koalas, that all koalas eat eucalyptus leaves, I make such a risky inference. I assume, among other things, that the koalas I observed are typical of their species. This assumption could easily be wrong, as I might have come across a peculiar band of koalas that happen to consume eucalyptus. That such inferences are risky, however, doesn't show that they are altogether unreasonable. The conditions under which they are reasonable are somewhat difficult to pin down, and we will tackle this challenge in the next section.

Now, given that reasoning nondeductively is risky, and that the conditions of its reasonableness are somewhat elusive, one might think that science should aim at just using deductive inferences, precisely because they are safe – even certain. But that would be a mistake. Remember: They are safe because in an important sense, they are uninformative. Since there is no new information in a deductive conclusion that was not already implicitly contained in the premises, deductive inferences won't allow you to gain more information about the world by reasoning from your evidence. To accomplish this, we need to go beyond an obsession with certainty, which is provided by conclusive reasons and reasoning, and enlarge our toolbox. The tools we need, especially for the empirical sciences, are various forms of *defeasible* reasoning, and thus *defeasible reasons*.

1.1.2 Defeasible Reasons

The second, and much more common, type of epistemic reasons are called *defeasible* reasons. They are also sometimes called *probable*, or *prima facie* reasons. The main difference between these and conclusive reasons is that even true defeasible reasons don't guarantee the truth of their target belief. Consider this example:

> You are near the mouth of a cave looking at a rock formation just inside the cave. The formation looks red to you. This "red-looking" is a good (defeasible) reason for believing that the rocks *are* red. However, as we all know, lighting conditions vary in natural settings and can be deceptive. Thus, it could be the case that the rock formation isn't really red; its red appearance could be produced by weird lighting filtering into the cave. Thus, although the red-appearance of the formation is a good defeasible reason for believing it to be red, the truth of this latter belief is not guaranteed.

What defeasible reasons do is to make the truth of the belief for which they are reasons *probable*. (That's why they are also called probable reasons.) A red appearance of a rock formation makes it more probable that the formation is red than that it is not. Of course, you could acquire a further bit of information which *defeats* the strength of the reason (that's why they are called defeasible reasons). For example, you could notice that there is a brilliant sunset outside the cave, which makes it likely that many even nonred things look red. Thus, the fact that the formation looks red to you is no longer a very good reason for believing that it is actually red, given that many nonred things will seem to be red in these lighting conditions.

There are actually two recognized kinds of defeating information, or defeaters: so-called *rebutting defeaters* and *undercutting defeaters*. In the example just given, noticing that there is a sunset is an *undercutting* defeater. It undercuts the evidential force of your original reason for believing that the formation is red. Given that you know of the red lighting, you now can't fully trust that things have the color which they seem to have. Of course, the rocks could still be red. But you would need to illuminate them with a white light or take a sample and observe it during daylight to make sure. On the other hand, it could also be the case that a geologist tells you that the formation isn't red, because there are no rocks of this color in the region. To the extent that you can trust her, you now have a *rebutting* defeater for your belief that the formation is red.

As mentioned above, defeasible reasons constitute the vast majority of the reasons we have for believing something. Conclusive reasons are limited to mathematics and logic. It is therefore extremely important to remember that talk of "physical proof," for example, is a misunderstanding of the concept, if "proof" is being used with its technical meaning. A proof consists in providing conclusive reasons for some target belief. That means that it must be literally impossible for the premises (i.e., reasons) to be true *and* the target belief to be false at the same time. Such high standards of evidence are unavailable in the empirical sciences. We just can't be certain. Even the best empirical evidence, on the basis of which we can form true premises for an argument, does not guarantee the truth of the target belief (i.e., hypothesis or theory; we'll explore the difference between these below).

Good evidence makes the supported theory highly probable, often so probable in fact that we may legitimately say that we justifiably assert that it is true. But the truth is not guaranteed by even the best evidence. Thus, there can't be literally a proof of any empirical belief – not the Big Bang, quantum mechanics, plate tectonics, or evolution. Proof requires conclusive reasons which are limited to mathematics and logic. This seemingly technical point is borne out by the history of science. Most, if not all, empirical theories of the past have turned out to be strictly speaking false, even though many were in the right ballpark and much better than the alternatives. Had the evidence used to support them guaranteed their truth, they couldn't have turned out false. But they did. Let's look at the infamous "substance" *caloric*, the stuff heat is allegedly made of – a once entirely reasonable and well-supported hypothesis that was ultimately wrong.

There was once good evidence for the theory that heat is a sort of subtle fluid, exchanged between bodies. This fluid was called "caloric." We still use the term "calories," which derives from that old theory of heat. It was partially based on the observation that if two objects at different temperatures are in contact, they will eventually reach thermal equilibrium: the hotter object will cool off and the cooler object will heat up, until they both have the same temperature. This looks a lot like what happens when one opens the valve of a hose connecting two buckets that hold different amounts of water. The water will run from the bucket with more water to the other one, until both hold the same amount of water. If heat was also a fluid, we would have a good explanation of the fact that adjacent bodies eventually reach thermal equilibrium. So based on this evidence and reasoning from what seemed to be an analogous system, scientists inferred that heat is such a fluid and called it caloric.

Of course, now we know that heat is, roughly speaking, micromotion or the movement of unobservably small particles (atoms and molecules). This understanding arose from the production of canons in the eighteenth century.[1] Using a cold drill to bore a hole into cold metal produced a lot of heat. But where was the heat coming from? Caloric was assumed to be a conserved quantity that could only be redistributed among objects, but neither created nor destroyed. The "new" heat from the canons was tantamount to running a hose between two half-full buckets and having them both overflow. You might think that this problem would've been apparent from other instances of friction, but the phenomenon was dramatically evident in the manufacturing of canons. What this episode shows is that even good evidence for a theory provides at best a defeasible reason for accepting that theory. New evidence can force us to retract our theory, exactly like knowledge of weird lighting conditions in a cave can force us to retract our belief about the color of rock formations.

1.2 Reasoning from Evidence

Scientists most often make use of *induction*, or drawing conclusions from evidence. Generally speaking, induction is *an inference from the observed to gain information about the unobserved (or unexamined)*, and it takes three different forms: *statistical inference, inductive generalization*, and *inference to the best explanation (IBE)*.

It is universally recognized that what these three forms of reasoning have in common is that they are defeasible. Beyond that, the terminology here is unfortunately not as widely agreed upon as it is with respect to deductive reasoning, but we can at least try to distinguish clearly between the three different forms of inductive reasoning just mentioned. Notice though that some textbooks in the sciences define inductive inferences as inferences from the particular to the general. This is misleading, because it covers only a tiny fraction of inductive inferences. We start with statistical inference.

1.2.1 Statistical Inference (SI)

The simplest example of an inductive inference is that of inferring something about an entire population from observing only some of its members. Recall the earlier example involving koalas: I inferred from having observed 20 of them munch exclusively on eucalyptus leaves that *all* members of the species *Phascolarctos cinereus* (that's the koala's scientific name) feed on eucalyptus leaves. Of course, such inferences are not restricted to biological populations. We might conclude that all igneous rocks are black, after we have seen many lava fields and observed that all of those were black. We can characterize the nature of statistical inferences in the following way:

> A statistical inference is an inference from the observed frequency of a property in a sample to the claim that the same frequency holds for the population from which the sample was taken, within a certain margin of error.

Here is an example in explicit form:

Premise 1:   The frequency of red marbles in a sample of 200 balls drawn from an urn was 49%.

Premise 2:   The urn contains exactly 1,000 marbles which are either red or black.

Conclusion: The frequency of red balls in the urn is 50%, with a margin of error of ± 2%.

Obviously, SI is an inference from the observed (the sample) to the unobserved (the population). Suppose you randomly picked up the first one hundred plants in a meadow and every one of them was a grass. You might well infer that every plant in the field was a grass. As we all know, beliefs (or hypotheses) based on SI can turn out false. Not all igneous rocks are black, and it's unlikely that all plants in a meadow are grasses, although koalas seem to invariably eat eucalyptus. Often, this is due to sampling problems, which can never be fully eliminated (maybe all the tall plants that are easily accessed are grasses, but some small, ground-hugging plants are broad-leaved species). But even if the sampling doesn't involve any bias, evidence from samples provides only defeasible reasons for beliefs about the relevant population, as the deviation of election results from predictions based on sampling (called polling) clearly demonstrates. There is much more to be said about SI, some of which you'll find in later chapters.

1.2.2 Inductive Generalization (IG)

This form of inference is a bit more difficult to characterize to any great degree of precision. In fact, not even the name is widely agreed upon. Sometimes, IG is used to refer to what we call SI. Since nomenclature is a matter of convention, nothing really turns on it, as long as we are reasonably clear about the differences among the kinds of inferences. In order to begin developing a good understanding for what we decided to call IG, it's best to start with an example.

Suppose you are interested in determining the functional relation between the period of a pendulum (how long it takes to pass through one cycle) and the length of its string. Dutifully, you plot changes in the dependent variable (the period) against variations in the independent variable (the length of the string). Unavoidably, you'll get a general trend with a somewhat messy point distribution. If you were to precisely connect all the points, you'd end up with a jittery line. "Nature can't be that crazy," you mutter to yourself, as you begin accepting that some of the points might not fall exactly on the line describing the actual relationship. You know about air resistance, the variable elasticity of the string due to changes in ambient humidity levels, the imprecision of your starting and stopping the timing device, and other factors that really have nothing to do with the true relation between length and period (that's why we call those factors "noise"). Thus, you decide to go for a nice, neat line – a section of a parabola, as it were. Then, you find the algebraic expression that generates that line. Finally, you make an inductive generalization and conclude that the period T of all pendula is related to the length l of their respective strings as follows: $T=2\pi\sqrt{l/g}$, where g is the gravitational acceleration. In fact, you are proud to have discovered the ideal pendulum *law*, which holds for all pendula with sufficiently small angular displacement. We will say more about the question of what a law of nature is in Chapter 12.

1.2.3 Inference to the Best Explanation (IBE)

This form of inference is exemplified by the story of caloric. The idea is that we observe some regularity and then postulate one or more mechanisms (or entities) that could be responsible for the observed regularity. If we can come up with just one, and it strikes us a plausible, we call it a day and accept it (this simplification will be corrected momentarily). If there is more than one, we look for the explanation which strikes us as best. To see what's going on here, look at the following example, inspired by philosopher Elliott Sober.[2]

Suppose you are sitting in your living room and suddenly hear strange noises coming from your attic – a quick succession of what sounds like little taps and then a rumbling noise. You consider two hypotheses: First, the noise is produced by gremlins from outer space that have landed on your roof and are now bowling in your attic. Call this hypothesis G. Second, the noise is produced by the neighbor's cat, which got into your attic and is trying to catch mice, but keeps running into the books you have stacked up there. Call this hypothesis C. Clearly, C is better than G, although G is an explanation of sorts: If it were true, then the probability of hearing those noises would be quite high. However, the same is true of C, and since C is more plausible than G in light of all the other things you believe, C is clearly the better explanation. You then infer that the best explanation is the most probable one and thus accept C. You have inferred the best of the explanations under consideration; this is IBE.

Several things are worth commenting on. First, both G and C make the observation probable – both bowling gremlins and mouse-chasing cats could produce those noises you hear. Second, both C and G might be false. Maybe it's neither cats nor gremlins, but it's some neighborhood kids playing a practical joke. Third, and related to the last point, what hypothesis counts as the best is partially determined by which ones you can come up with. In other words, the best hypothesis from among those we thought of need not be a very good one, all things considered. Often, we miss an even better hypothesis, as we know from the history of science and discuss further in Chapter 12. In fact, the history of scientific progress is one of not only gathering more evidence but improving explanations such that was once the "best" is supplanted by something better.

Thus, even if you are very creative in generating hypotheses, you might generate really awful ones and shouldn't believe any one of those. This has prompted some to eschew the use of IBE altogether, especially insofar as it pertains to unobservables (things we can't directly see or otherwise sense, such as electrons or magnetic forces). In short, IBE can provide some reason for accepting a claim (we use it in forensic sciences all the time, for example, when we try to find the person whose presence is the best explanation of all the clues, and infer that the person who best fits the clues is the perpetrator), but it certainly doesn't guarantee our knowing the truth.

Finally, it is important to point out that IBE cannot be reduced to other forms of inductive inference. Inferring the presence of a stray cat in my attic as the best explanation of the noise I am hearing does not (need to) involve prior observations of stray cats *in my attic* and their behavior. Thus, this inference is different from a statistical inference, which, if you recall the example of koalas, does rely on observations of the feeding habits of a number of koalas to infer something about what other koalas will eat. Neither am I trying to establish any sort of regularity when I infer that a cat must have gotten into my attic.

I am simply interested in explaining this particular and odd event by evaluating various hypotheses as to their plausibility in light of my background knowledge about cats, none of which I need to have oberved in an attic.

Let's recap what we discussed in this section. We distinguished between conclusive and defeasible reasons. Along with that distinction comes another one, namely, the distinction between deductive and inductive reasoning. The former can be characterized as an inference from the implicit to the explicit, while the latter is an inference from the observed to the unobserved. The following diagram (Figure 1.1) shows the distinctions you should keep in mind:[3]

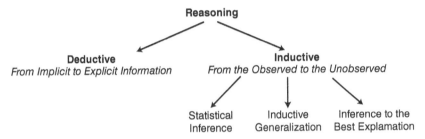

Figure 1.1 Forms of reasoning.

1.3 Knowledge and Truth

Scientific inquiry is often characterized as an especially promising way to increase our knowledge of the world by discovering more and more truths about it. But what exactly is knowledge, and what is truth? To answer the first question, we can start by determining what the difference between mere belief and knowledge is. Suppose Marvin strongly believes that atoms are held together with hooks. Does Marvin know this to be the case? It seems not, for the simple reason that it is false. We can't know what's false. Of course, we can believe what's false, and we can falsely believe that we know something even though it is in fact false. But we can't know that $2 + 3 = 78$. Thus, we can conclude that in contrast to mere belief, a belief that counts as knowledge has to be true.

But being true is not enough for a belief to count as knowledge. Suppose I am terrified of snakes, flipping a coin every morning before entering the lab to see whether I have to use protective gear. This morning, the coin landed heads up, which means, to me, that there is a snake in the lab. Moreover, there in fact is a snake – unnoticed by me, a practical joker snuck it into the building. Do I know that there is a snake? It seems not. That the coin landed heads up doesn't provide a good reason for believing that there is a snake. And yet it is true. This shows that for knowledge, we don't just need truth – we need good reasons as well.

Putting it all together, knowledge amounts to true beliefs that are based on good epistemic reasons. Some of the readers might be aware of the fact that over the last several decades, it has become clear that even true beliefs for which I have good reasons might not automatically qualify as knowledge – something else is needed, at least in the kind of instances usually called *Gettier* cases.[4] However, because we encounter such cases only rarely, if at all, during scientific research, we won't discuss them here. In the next couple of chapters, we will instead explore the notion of evidence and how it is supposed to yield good reasons for our beliefs about the world, thus providing us with one ingredient of scientific knowledge.

The second ingredient is truth. There are many competing theories of what truth is, and there is an important debate in the philosophy of science whether or not our best theories are true. We can't explore the first question here, but we will explore the debate about the alleged truth of scientific theories in Chapter 12. For now, we can just stipulate that by a true theory, or a true belief for that matter, we mean a theory (or a belief) that corresponds to the facts in the world. For example, the hypothesis that you are reading this sentence right now is true because it corresponds to the fact that you are reading it.

1.4 Facts, Hypotheses, and Theories

This may be a good point for some remarks about the relation between facts, hypotheses, and theories. To begin with, facts are actual states of affairs in the world, or how things are. For example, there is the fact that you are reading this sentence right now. This is something that is happening in the world – you reading this sentence is among all the facts that together determine the world. In contrast, hypotheses and theories are (often) linguistic expressions of our beliefs: They consist of sentences purporting to report the facts. For example, the sentence "You are reading this chapter right now" is a hypothesis about you and what you are doing. The hypothesis is right now true (you are still reading, aren't you?), but once you get bored and look for your friends on social media, it will become false. And this happens because the facts will have changed.

The difference between fact and either theory or hypothesis can be easily overlooked. For example, in discussions about evolutionary biology and creationism, one often hears creationists complain that *evolution is just a hypothesis*, while the evolutionists call the *theory of evolution* a *proven fact*. Both sides here are equally guilty of speaking carelessly. First, *evolution* is not a hypothesis, because evolution is not a linguistic entity. It is a process that occurs in nature. *Evolutionary theory*, on the other hand, is a linguistic entity (it's a bunch of sentences written in books), which can be either true or false. And, of course, evolutionary theory is true if and only if (iff) it corresponds to the facts that comprise the process of evolution, such as the facts concerning natural selection.

Second, there are no such things as "proven facts." As we have already seen, only mathematical or logical sentences (and the propositions expressed by them) can be genuinely *proven*. Empirical theories can only be justified by defeasible reasons, which, however, are often strong enough to warrant treating those theories as reporting the facts correctly. Moreover, facts are not the sort of thing that can be proven. You can't deduce a fact from some premises, because they are not linguistic entities. What you can do is try to bring about some fact, by acting in various ways (e.g., you could apply antibiotics to bacteria to bring about the evolution of resistance). Or you can discover facts. But you can't prove them – they are the wrong kind of thing for that.

Finally, what's the relation between hypotheses and theories? It seems that we often try to use the distinction to mark a difference in how well supported an empirical claim is and how comprehensive its content is. We demand better support for a theory than may be required for a (mere) hypothesis, which is often thought of as something like an educated guess, a conjecture, or a sort of hunch.

If we approach the matter a bit more systematically, we can think of hypotheses as relatively isolated or "stand alone" empirical claims. For example, in scientific modelling, we are often only interested in correctly identifying relations between dependent and independent variables (such as period and length in a pendulum). This can be done regardless of how the resulting model fits together with other things we believe.

When we propose a theory, however, we typically try to fit several hypotheses together into a coherent whole. Recall the earlier example involving the ideal pendulum law. During the process of fitting a curve to the data points you collected, you might think of the result as your hypothesis about how period and length are related. You can then go on and embed this hypothesis into a comprehensive theory or encompassing model, such as the theory of simple harmonic oscillators, or even more broadly, into all of classical mechanics.

"Embedding" means that you try to show how your hypothesis can be derived as a special case of a simple harmonic oscillator or as an instance of Newton's laws of motion (together with some boundary conditions). Such an embedding also means that the evidence for the other parts of the theory becomes indirect evidence for your original model.

Thus, we can draw the distinction between hypothesis and theory in terms of how many different pieces of information have been integrated, although we have to keep in mind that this distinction is vague.

1.4.1 "It's True for You but Not for Me"

One topic that often comes up in discussions about truth – a topic we haven't touched on yet – is that of *relativism*. Many people profess to be relativists about truth. They might say things such as: "This is true for you but not for me," or "It was true for Ptolemy that the earth is stationary, but now it is no longer true." Some might even provide a theoretical defense of the view that truth is relative to one's culture.

What is often in the background here is the idea that one needs to respect another person's opinion and perspective. After all, wars have been fought over differences in beliefs. So epistemic humility is a virtue. But this gives rise to a problem. If I disagree with another person, I seemingly need to hold that truth is a matter of opinion, perspective, or culture. However, there is an alternative.

One can respect other people's opinion and perspective in that one can appreciate how their evidence and cultural experiences might make it seem that so-and-so is the case, while my evidence makes it seem that this is not so. Remember, the defeasibility of our reasons is a sound basis for doubt. The evidence people have access to often varies dramatically among individuals.

It might even be the case that individuals have access to the same body of evidence but give different weights to the various elements based on experience, existing beliefs, practical interests, and cultural values. We'll explore this possibility in some detail in the next chapter, but for now let's just say that once we establish the spatiotemporal context (and sometimes the values), we often find that there is little disagreement. Once again, in philosophy as in science, one must take care to make matters clear and, in this case, to not mistake differing interests or perspectives for there being relativistic truth.

Thus, what it is reasonable to believe can vary dramatically from person to person. We have already seen earlier that it can be quite reasonable to believe something that's false. The evidence one has might in fact not be very good evidence. Thus, I can respect opinions that are different from mine for their reasonableness without having to claim that truth varies from person to person. Once we make a distinction between two ways in which beliefs can be "good" (i.e., their reasonableness and their truth), we can explain variations among what reasonable people believe in terms of differences in evidence without having to be relativists about truth.

And that is a good thing, because relativism about truth is ultimately indefensible. Take, for example, the sentence, "There is no objective truth." Is that sentence itself true? If it is objectively true, then it has just refuted itself. If it is merely relativistically true, then why should I, who believes in objective truth, be moved by it? Subscribing to relativism about truth puts one into a really bad intellectual (and practical!) position.

Of course, these remarks are far from sufficient for putting relativism to rest. There are fairly sophisticated defenses of this position, but it would lead us too far into epistemology proper to give them a fair hearing. What the remarks do show, however, is that relativism as a knee-jerk reaction to the existence of disagreements about truth is simply too naïve.

1.4.2 Perspectivism

Although relativism, at least in its unsophisticated form, is not a plausible response to disagreements about what is true, there is a viable approach that allows for the truth of multiple claims about the world, but not the truth of every claim as relativists would contend. Ronald Giere advocated for *scientific perspectivism*, a philosophical position mediating between the hard-driving objectivism of most scientists (as well as the unyielding realism of many philosophers of science), and the "anything goes" constructivism found among postmodern social scientists and humanists. Perspectivism contends that while there is an objective reality, our statements about its features are always made from a particular viewpoint (both literally and figuratively) and are hence partial truths, at best. This position fundamentally differs from relativism in that once a perspective is chosen, claims about the natural world can be empirically tested and refuted or confirmed.

The notion that truth claims are always relative to a perspective is, as Giere points out, not radical. Even the staunchest empiricists cannot escape the fact that scientific claims are relative to language. What Giere initially proposed was to extend linguistic contextualization to one's physical position (both in relation to the object of study and the internal processing of the incoming signal).

Consider, for example, a biologist who asserts that the petals of a particular flower are white. One need only look to see that this is true. However, suppose another biologist comes along with a device which reveals ultraviolet reflectance, looks through this instrument at the flower, and declares that it is striped. And, in fact, some flowers have so-called "bee guides" that direct bees, which can see in the ultraviolet, to the stamen and nectary which provide pollen and nectar. Which description of the flower is true? The perspectivist would respond that the truth of the competing claims depends on whether the accounts are rendered from the viewpoint of a human or a bee. Note however, that once the perspective is fixed, the descriptions can be wrong; from the human perspective it would be mistaken to call the flower red, and from the bee perspective it would be wrong to assert the flower is checkered.

While this example illustrates that scientific claims are conditional on biophysical and instrumental perspectives, Giere also maintained that propositions are made within a theoretical context. For example, a sociobiologist might contend that helping a neighbor is explained in evolutionary terms by reciprocal altruism, while a cultural anthropologist might argue that cooperation serves to strengthen social bonds in a community. Along similar lines, a neurophysiologist might explain that people eat sweet foods because sugars stimulate nerves leading to the gustatory and reward regions of the brain, while an evolutionary biologist might explain that sweetness is favored by its association with concentrated sources of energy which were rare in the history of our species. Giere maintained that extreme pluralism arises from a multitude of competing, theoretical perspectives in empirically weak scientific fields at the limits of experimental capabilities (e.g., gravity waves).

Others philosophers have pushed perspectivism further, to take into account the possibility of different interests. That is, to assess a scientific claim, the needs and desires of the investigator are relevant. When asked if a forest fire should be allowed to burn, an ecologist might answer that the dominant trees in the ecosystem of concern are pyrophytic, requiring fire to reproduce, so "let it burn." But another ecologist might answer that the last surviving population of the endangered yellow-spotted salamander lives in the ponds within the burning forest and the fire should be extinguished to conserve this species. And an ecologist working for a forest products company might insist that jobs would be lost if the fire is not controlled.

You can see that perspectivism admits that scientists may have differing values and, in fact, there is no value-free science (at a minimum, scientists value knowledge). If the forest fire is allowed to burn it is true that the plants will recover (a long-term perspective valuing natural processes), that the amphibian will go extinct (a perspective valuing the existence of species), and that people will be harmed (a short-term perspective valuing economic well-being). So, while logically incommensurable claims are not countenanced from a given perspective (e.g., it cannot be the case that the salamander is both endangered and flourishing at this time), incommensurable conclusions can be drawn from scientific data when perspectives diverge (e.g., forest health might be harmed in the short-term and benefited in the long-term by the same event).

1.5 Conclusion

This completes our brief survey of some central concepts in epistemology, such as the concepts of epistemic reasons, knowledge, and truth. As you can imagine, there is much more to be said on all of the topics we touched on. The suggested readings provide further and deeper treatments of many of those topics, if you are interested. Having shown that scientific knowledge most often involves some form of inductive reasoning from evidence, we next go one step deeper and turn to a fundamental question in the philosophy of science: What is the nature of evidence, and how can we access evidence?

Notes

1 The debate between a dynamic theory of heat and the defenders of the caloric account was an important stage in the development of thermodynamics. Early defenders of the dynamic theory included Galilei Galileo, Isaac Newton, and Robert Boyle, who thought that heat is due to the internal motion of particles. This however was difficult to reconcile with the discovery of latent heat (i.e., the heat absorbed by a substance that undergoes phase change, e.g., melting, but doesn't rise in temperature) in 1757 by Joseph Black. For details, see Robert D. Purrington, Physics in the Nineteenth Century, Rutgers University Press, 1997, pp. 75–101.

2 Elliott Sober, Philosophy of Biology. Westview Press, 1993, p. 32.
3 A form of inference related to inductive inferences is reasoning by analogy. We'll say more about this in Chapter 9.
4 A very simple Gettier case involves a clock whose hand is stuck at 5. If you don't know the clock is broken, you are justified in believing the time it indicates (it has worked reliably so far). Moreover, if you just happen to look at that clock when in fact it is 5 o'clock, you have a true belief that is also justified. But few would say that this justified true belief counts as knowledge of the time, because your belief is only accidentally true.

Annotated Bibliography

Ronald Giere, 2006, *Scientific Perspectivism*. Chicago: University of Chicago Press.

Giere argues that scientific claims are conditioned on highly confirmed theory or reliable observations such that if the assumptions of the underlying theories or the biophysical properties of the instrument/observer changes, then the result can be a different, even incommensurable, accounts of the natural world. There is no complete, objective, or singularly correct account of the world through correspondence how the world really is.

Alvin I. Goldman and Matthew McGrath, 2015, *Epistemology. A Contemporary Introduction*. New York and Oxford: Oxford University Press.

This book contains advanced discussions of the structure and nature of justification, the Gettier problem, skepticism, and contextualism. It also covers methodological questions concerning the use of intuitions as evidence as well as important new developments in social epistemology which are important for understanding how knowledge emerges from research teams, as opposed to individual scientists.

David Manley, 2019, *Reason Better: An Interdisciplinary Guide to Critical Thinking* (Tophat Monocle) https://app.tophat.com/e/455176/assigned

This recent and very accessible book on critical thinking covers many of the concepts we encountered in this chapter, such as reasoning and inference. In addition, it contains illuminating discussions of various biases, inferential fallacies, and cognitive illusions that can interfere with successful reasoning.

2

EVIDENCE, OBSERVATION, AND MEASUREMENT

2.1 The Promises of Evidence

What's so great about evidence? Science is often regarded as the paradigmatically rational way to gain knowledge about the natural world. In particular, it has been argued that in contrast to ideology, religion, and mysticism, science is both appropriately responsive to how the world actually is *and* allows for a rational resolution of disputes between different theories about the world. Both of these features of science are intimately tied to its use of *evidence*. Evidence, at least ideally, is publicly accessible, sharable, and can serve as an arbiter in the competition between different theories. In other words, in contrast to disputes between different religious beliefs, political worldviews or economic ideologies, which typically cannot be resolved by appealing to a neutral standard, science solves such disputes in a rational manner, because it *has* access to a neutral standard: evidence.

Let's look at this in more detail. Generally speaking, evidence typically *indicates* (although we'll see important exceptions) particular facts about the world. Tracks in the snow, for example, may indicate the prior presence of a mule deer, and those tracks are thereby evidence for the mule deer's earlier presence in the area. How do those tracks provide evidence for that fact? The answer is straightforward: The deer *caused* the tracks that you are seeing in the snow. This means that at least this one form of evidence is grounded in *causal relations* between different features of the world. In our case, the causal relation was one between the deer's hoofs and the tracks. And there are numerous similar cases: The victim's blood on John's shirt is evidence for the hypothesis that John was at the murder scene, because the injuries of the

This is Philosophy of Science: An Introduction, First Edition. Franz-Peter Griesmaier and Jeffrey A. Lockwood.
© 2022 John Wiley & Sons, Inc. Published 2022 by John Wiley & Sons, Inc.

victim caused her blood to end up on John's shirt. The so-called Koplik spots on a person's gums are evidence for the hypothesis (often called a diagnosis in this context) that the person has the measles, because having the measles causes those spots to appear. In fact, there are so many cases in which we have observable indicators (i.e., evidence) for features of the world we might not directly observe (deer, measles, etc.) that one might be tempted to simply define evidence in terms of a causal indicator relation, maybe along the following lines:

> *An observable feature O of the world is evidence for the existence of an unobserved or unobservable feature U of the world if and only if U caused O, and O thereby indicates the existence of U.*

A brief remark on the distinction between unobserved and unobservable features, invoked above, is in order. In our first example, the deer was simply unobserved, but is of course observable. But there also seem to be clear cases where we can't observe that for which we might have observable evidence. Newtonian gravitation is a good example. We can observe its effects when we see a football drop into the outstretched arms of the receiver. However, we can't observe the gravitational force directly. In fact, it seems we can't observe any of the fundamental forces of nature directly. For this reason, many have argued that forces are prime examples of entities that are in principle unobservable, and that the only reason we have for believing that they exist at all is the evidence manifest in their actions. We can see neither the strong nor the weak nuclear force, but we can see their effects, the most prominent of which are of course all those familiar physical objects that surround us.[1] This is the reason that the proposed analysis of the notion of evidence refers to both unobserved and unobservable features.

Let's go back now to the idea that evidence can be defined in terms of causal relations between different features of the world. If this idea is correct, it would also explain nicely why scientific disputes can be resolved by appealing to a neutral, objective standard, as we claimed earlier. At first glance, it seems that causal relations are objective features of the world. Their existence is independent from human observers – the extinction of the dinosaurs was caused by certain features of the world (most likely by a meteor that hit the earth), even though there were no humans to observe it. Thus, what humans believe has no influence on the causal relations that exist in the world. This explains how evidence can help us decide between competing theories: Since two genuinely differen theories will not agree on all possible evidence, then, if we find evidence that can be better accounted for by Jane's theory than by Bob's, we have objective reasons for deciding the competition in favor of Jane. The reasons are objective, because causal relations that ground evidential relations are objective. Thus, if Bob holds on to his theory despite the evidence favoring Jane's, he acts unreasonably.

This last observation leads to a further important fact concerning our ordinary concept of evidence. We think that our beliefs about the world have to be properly responsive to empirical evidence, and if they are not, those beliefs appear irrational. Thus, there is an important connection between being rational and being responsive to evidence. This connection will be the topic of later chapters. For now, our next task is to discuss what items in the "world" are appropriate candidates for evidence.

2.2 Basic Evidence and Derived Evidence

2.2.1 What We See

One question that many philosophers, as well as some scientists, have asked (at least implicitly) is this: What should be regarded as ultimate or basic evidence? In other words, what is our evidential bedrock? This question might seem strange at first. Clearly, it is information about the world and what is happening in it that should serve as the ultimate evidence to be used in the construction and confirmation of our empirical theories. For example, a physician's observation that her patient exhibits Koplik spots is (a piece of) the ultimate evidence for her diagnosis that the patient has the measles. Or, to take another example, my gas gauge's needle position is my evidence for the belief that I have enough gas in the tank to get home easily. In light of these examples, it would seem that ordinary observations of the world, along with measurements, constitute our basic evidence.

However, things are not altogether that easy. Consider the expression "information about the world" more closely. In what form do we access this information, and how do we process it? One might think that the observation of, say, Koplik spots is unproblematic. Looking at her patient's gums, the physician simply observes the presence of those spots. But is that really the best way to describe the situation? That it might not be becomes clear when we imagine a person who has never heard of Koplik spots, or of the measles, and who has never seen those "spots" before. Clearly, to him, what he observes are some lesions or blisters on the gums, but not Koplik spots. A lot of theoretical knowledge goes into observing Koplik spots as opposed to observing mere discolored tissue. Pushing this line of thought even further, an extraterrestrial life form, for example, one who hasn't seen gums, or humans, before, won't even see spots of tissue discoloration. It might describe its observation as that of a reddish surface with small whitish regions, if indeed it has the concepts of "whitish" and "reddish." What this seems to show is that what we directly observe are not things of a certain kind (such as human gums) carrying indicators of a disease (such as Koplik spots), but rather, what we directly observe are distributions of colors and edges in our visual field.

The same point is perhaps even more obvious in the example involving my gas gauge. Clearly, I need to know what the instrument is supposed to show before I can use the needle position as evidence for any hypotheses about the amount of gas left in my car, and thus as evidence for being able to get home safely. (Many of us have probably experienced confusion about a brand-new car's instrumentation.) Furthermore, I already need to know what roles needles play in an instrument, and even what instruments in general are, viz., measuring devices of various kinds. None of this information can be gleaned directly from what we actually observe: distributions of colors, surfaces, and the like.

Considerations such as these have led many to propose that the ultimate, or basic, evidence for all of our hypotheses about the world are various distributions of colors, edges, sounds, and the like, which result from impacts of the world on our sense organs. As the American philosopher Willard van Orman Quine famously put it:

> "[…] the only information that can reach our sensory surfaces from external objects must be limited to two-dimensional optical projections and various impacts of air waves and some gaseous reactions in the nasal passages and a few kindred odds and ends. How […] could one hope to find out about that external world from such meager traces?"
>
> (*Roots of Reference*, 1974, 2)

Quine raises two issues here. First, he claims that our contact with the world is constituted by the impact of various forms of information, embodied in various physical systems, on our sense organs. Second, he rightly notices that it is somewhat surprising that we can have knowledge of the world on the basis of such "meager traces." We won't discuss the second issue now, which is related to the question of skepticism and realism, but relegate it to Chapter 12. Instead, we will briefly discuss the view that the ultimate evidence is to be found in distributions of various physical properties and/or magnitudes in one's sensory field. The result of our discussion will be to dismiss this view as a fruitful account of scientific evidence. We will also see that the view possibly arises from equating, rightly or wrongly, evidential with causal relations.

2.2.2 Causes and Evidence

The main issue with the view that the ultimate evidence consists in features within a person's sensory field is that it seems highly unnatural. Clearly, I don't seem to observe edges and colors and only then form, by some unknown process, beliefs about my environment. I don't seem to *first* see rectangularly shaped red patches supported by a shiny, gray surface and *then* infer that there are books with red covers in my old metal bookcase. I seem to see the books and the bookcase themselves, directly, as it were. If someone were to ask me why I believe that my books are still in my bookcase, I would certainly say something like, "Because I can see them," rather than, "Because this particular distribution of red rectangles on a shiny gray surface is good evidence for the presence of my books in the bookcase." And yet, it is of course correct to say that seeing the red rectangles on a gray surface is involved in my belief that the books are still there.

In order to resolve this tension, we need to make a distinction between a description of the processes by which we become aware of our environment and the question as to what counts as evidence for what. Cognitive science tells us that our beliefs about the world stem from stimulations of various sense organs, which are then processed by our brain and result eventually in beliefs about what's out there. However, this process from sensory stimulations to beliefs is not obviously the same as the grounding of our beliefs on evidence. Reflecting the ordinary ways in which we talk about evidence, we will adhere for our purposes to a distinction between causal and evidential relations. The path from sensory stimulations to beliefs about the world consists in a long chain of causal links. To correctly describe and understand this causal chain is the task of cognitive science. The task of the philosophy of science is to isolate those elements in the causal chain that stand in an evidential relation to the relevant beliefs.

What is the evidential relation? For our purposes, it may be enough to say that some observation counts as evidence for a person's belief if the observation provides the person with a good reason for her belief. Seeing books in my book case provides me with a good reason for believing that my books are there. Ordinarily, I do not even become aware that the observation of my books is based on particular sensory stimulations. Thus, those sensory stimulations, in virtue of going on outside of my awareness or unconsciously, do not provide me with good reasons for believing my books are still in their bookcase. The sensory stimulations are the ultimate cause of my seeing and hence my belief, but since they do not provide me with a good reason to have the belief, they do not count as my evidence.

Of course, what counts as a good reason for a person to believe something depends partly on factors involving that person, most importantly perhaps their background knowledge. Recall the aliens from earlier. Not having any knowledge about humans and their diseases in the background, the evidence in the form of Koplik spots is not available to them. Just as we argued that if I am unaware of my sensory stimulations, they can't be reasons for me, if the aliens don't know what Koplik spots are and what they indicate, the spots do not provide good reasons for them to believe that the person in front of them is about to come down with the measles. But we don't have to invoke aliens. Any nonexpert looking at Koplik spots will fail to receive evidence for the measles.

Thus, whether or not an observation is evidence for a person depends crucially on what the person is bringing to the table in terms of background knowledge. The physician sees the Koplik spots *as* Koplik spots and thereby gains evidence for her belief that the patient is coming down with the measles. I, a nonexpert, see the same spots, but I do not see them *as* Koplik spots. Thus, they are not evidence for me, because they don't provide me with good reasons for believing that I am confronted with a case of the measles. Only observations that are seen *as* this or that can be evidence for (or against) some hypothesis.

Another way of putting the same point is this: During the processing of sensory stimuli, we bring – often automatically – various categories, background knowledge, and similar things, to bear. Thus, categorized observations are what constitutes evidence. This fact has been identified as the *theory-ladenness of observation and/or measurement*, which we discuss below and in later chapters.

Some readers might worry that we are overintellectualizing evidence, seemingly requiring that evidence is always conceptualized (e.g., seeing a discoloration *as* a Koplik spot). Such readers are in good company – there are a number of prominent epistemologists who believe that nonconceptualized experiences can be good reasons for having certain beliefs. For example, simply tasting something salty is a good reason for believing it to be salty, even if we don't have the concepts to properly express our belief (that's why we often say that something tastes like chicken if we lack the appropriate concepts). However, it seems to us that in the empirical sciences, unconceptualized experiences hardly ever play a role as evidence, and so we leave them aside.

2.2.3 Observation, Naked and Enhanced

Another feature about observations is that when unaided, they can be both restricted in their scope and sometimes unreliable. Not only do we bring background knowledge to the categorization of our observations, but what we can observe in the first place is limited to and characterized by the way in which our brains process incoming information. First, there is a lot of information available in the world that our sensory apparatus cannot pick up directly. We can only see a small part of the electromagnetic spectrum – the wavelength of visible light stretches from about 390 to 700 nanometers. We obviously can't see infrared (however, some snakes can), and we can't see radiowaves or gamma rays. All of these waves can carry information. In order to tap into that information, we need special equipment, such as night vision goggles that allow us to "see" infrared, or radio receivers that allow us to hear the information superimposed onto the radiowaves.

Moreover, observing something with our unaided senses can sometimes lead to false, or inconsistent, beliefs. Consider trying to measure the temperature of an object with your hand. As it happens, you are coming home from a walk on a very cold afternoon. Somewhere along the way, you lost one of your gloves, so that only one of your hands was protected and is still nice and warm. The other hand has gotten quite cold. After you open the door, you are wondering whether the pot of soup you made earlier is still warm. You touch the pot with your cold hand and determine – observe – it to be still really hot. So you take off the glove from the other hand to pick up the pot and carry it to the table. Surprise! To the other hand, the temperature of the soup is barely above the temperature of the room. What's happening is that the temperature information you receive from your two hands does not only depend on the temperature of the soup pot, but also on the temperature of your hands. That's why you get inconsistent information. Hands, or thermoreceptors in the skin in general, are not designed to provide us with objective temperature readings; what they do is provide us with information about the temperature difference between environment and body.

In order to overcome these two shortcomings of unaided observations, we can – and do – avail ourselves of instruments that can both broaden the range of the information to which we have access and make the information less susceptible to distortions stemming from our sensory apparatus. Consider temperature again. We build thermometers that can measure the temperature of objects that would be far too hot or too cold to be touched by us without injury – molten iron, for example, or liquid nitrogen. In this way, we can get information we couldn't receive through unaided observation. Second, thermometers, like all other measuring instruments, are calibrated. We use certain natural facts, such as the phase changes of water, to equip them with a scale that allows us to map numbers on the thermometer to temperatures in the objects measured. In the Celcius scale, the freezing point of water is set to 0, and its boiling point (at sea level) to 100 degrees. This has the effect that we measure the temperature of objects by comparing them to the temperature of water (at normal pressure), instead of simply determining the temperature difference between the measuring instrument and its environment, as we do when we ascertain temperature by means of our own body. Thus, we eliminate some of the subjective elements in our observations. But, as we will see shortly, a different sort of bias manifests itself in the use of measuring (and other scientific) instruments.

2.3 Measurement

As we just noted, instruments for measurement are used for principally two reasons. First, they expand the range of information about the world available to us, and second, they set up a correspondence between numerical values and properties. The second feature in particular has received attention from those concerned about the subjective character of ordinary experience and the desire for scientific objectivity. If, so they thought, we could compare properties of different objects (taken in a wide sense) with a standardized measuring instrument as opposed to using our unaided observation, certain errors could be minimized. For example, trying to gauge the temperature of a pot with a very warm hand might lead to questionable results. Better to use a thermometer. However, there are some interesting philosophical issues that are raised by the employment of measurement, and a couple of those issues rise to the status of serious problems and puzzles, such as the matter of measurement in quantum mechanics. But we start with its less problematic features.

2.3.1 Measurement Scales

In general, a measuring instrument sets up a correspondence between numbers and physical properties. However, not all features of the number system (the natural numbers in most cases) are relevant to all measuring devices. What is relevant depends on the sort of scale one uses. We can distinguish between at least

- Ordinal Scales (e.g., hardness)
- Interval Scales (e.g., temperature)
- Ratio Scales (e.g., length)

First, *ordinal scales*. For minerals, the so-called Mohs Scale is the standard way for comparing their hardness. It consists of numbers 1 to 10 with which various minerals are associated. For example, in a standard Mohs laboratory hardness kit, we find 1 = talc, 2 = gypsum, 3 = calcite … 9 = corundum, and (if you have enough research money, you include) 10 = diamond. We measure the hardness of some mineral M by a scratch test. If M can be scratched by calcite but not by gypsum, and if M cannot scratch gypsum, we assign 2 as the number of its hardness. If, however, gypsum cannot scratch M, but M can scratch gypsum (but not calcite), we assign 2.5. In this way, each mineral specimen can be located on the Mohs scale.

Notice what the scale does not provide. First, it does not provide a unique number to each mineral. The scratch test cannot deliver verdicts for minerals that do not scratch each other except that they are of the same hardness. Second, and more importantly, it only delivers ordinal information, i.e., information about the location of the mineral on the scale. It does not, however, let you infer that if one mineral has hardness 2 and the other hardness 4, that the latter is twice as hard as the former. Diamonds are not, in any meaningful way, ten times as hard as talc. Finally, it does not even tell you that if M is 4, another sample N is 6, and still another one P, is 8, that P is harder than N by the same amount that N is harder than M. The Mohs scale does not provide this information, and it would be a mistake to believe that it does.

This latter sort of information can be gleaned from *interval scales*, however. Those are used, for example, for measuring temperature. The difference between 90 and 100 degrees F is the same as the difference between 80 and 90 degrees. With an interval scale, we not only rank temperatures on an ordinal scale, we can also measure temperature differences. What we still can't do, however, is make magnitude comparisons between temperatures. For example, something with a temperature of 60 degrees F is not twice as hot as something with a temperature of 30 degrees F. This is so because the Fahrenheit scale has no natural zero point. 0 degrees F does not mean the absence of temperature altogether (as it does with the Kelvin scale).

Contrast this with *ratio scales*, such as the centimeter scale. A measurement of 0 centimeters *means* the absence of extension (in the direction measured; never mind how this could be done). Thus, if you measure something to be

200 centimeters long, you *can* infer that it is twice as long as the thing which I measure to be 100 centimeters long. Therefore, the use of ratio scales provides more information than that of interval scales, and much more than we can get from ordinal scales: you can rank different objects by length ordinally, and you can measure differences in length. In addition, you can measure length ratios (thus the name). The interesting point here is that what information you can glean from the world is not only dependent on the physical characteristics of the measuring instruments, but also on the scale you choose to implement. Such choices are often determined by practical concerns.

Given that ratio scales deliver much more information than, say, ordinal scales, the question arises why we don't use ratio scales for measuring any property. The example of the property of hardness makes this implausible. What would it mean to measure hardness on a ratio scale? First, we would need an absolute zero point – but what material would exhibit the property of the absence of hardness? Second, even interval scales seem to be inapplicable to hardness. In what exact sense could differences in hardness be exactly same for different pairs of materials? What the implausibility of developing ratio (and interval) scales for some properties tells us about these properties is unclear. One possibility is that being hard is simply not a fundamental property of the universe, and that only fundamental properties are amenable to being measured on a ratio scale. However, it would be quite surprising if it turned out that the notion of a *fundamental property* corresponds to what scales are appropriate for its measurement.

2.3.2 Operationalism

An entirely different answer to the puzzle about hardness (and related ones) is that the measurement procedure actually defines our scientific terms in the first place. This view is called *operationalism*. According to it, measurement procedures give meaning to otherwise ill-defined predicates (those are the terms used to refer to properties). A famous example is the predicate "intelligent." An operationalist would claim that the term "intelligent," as ordinarily used, is unsuited for scientific purposes. We simply don't know whether or not it applies to the behavior of a person, because we don't have sufficient agreement between researchers as to what property is singled out by the term. Thus, we define "intelligent" more precisely as standing for a property which is causally responsible for the performance of a person in a standardized intelligence test. Such tests deliver numbers on a scale. The view is that those numbers may or may not measure any property we were interested in before we started to theorize, but rather single out a property as corresponding to a precise predicate that is scientifically useful. On this view, hardness may be a pretheoretical property, corresponding to different human experiences of resistance to pressure, but the "scientific hardness" of a sample of material is its location on the Mohs scale. End of story.

A form of operationalism was influential for Swiss physicist Alfred Einstein's development of the special theory of relativity; he also insisted that legitimate scientific concepts have to be defined by appropriate measurement operations. Thus, the meaning of "space" is given by the operations we apply for measuring length, and that of "time" by the operations for measuring duration. We will revisit this issue in Chapter 11. However, as a general theory of the meaning of scientific terms, operationalism is inadequate, as can be shown by the following consideration.

First, if methods of measurement indeed fix the meaning of a predicate, then there can be no such thing as an incorrect method of measurement, because there is simply no room for a mismatch between a property and how it is being measured – the former is defined in terms of the latter! For example, if intelligence is defined in terms of the standardized test with which we measure it, the test cannot fail to measure intelligence correctly. The measurement methods are correct as a matter of convention. But, second, it seems that it is possible to develop methods for measuring some quantity that are wildly inadequate. For example, measuring hardness by shining lights of various colors onto minerals seems clearly wrong, even if we supply a "scale" (looking prettiest under color 1 (red) = hardness 1). Seeing that this "measurement" of hardness is just crazy clearly shows that there is a well-(enough)-understood notion of hardness even before we try to measure it.

This seems to support a sort of realism about the properties we try to measure, in the sense that measurement is the estimation of the magnitude of an independently existing property. In particular with respect to properties such as length, weight, and duration, for example, a realist stance seems most plausible. That is, there are real, observer-independent features of the world. It certainly seems to be the case that objects have a determinate length independently from our attempts to measure it, and processes have an equally determinate duration. But as we will see later, things are actually not all that clear cut. It turns out that what length we measure depends on the motion of the object relative to the observer. This phenomenon is called length (or Lorenz) contraction, and we discuss it in some detail in Chapter 11. However, even with such simple properties such as mass, we can already notice that its measurement is at least in some cases heavily influenced by already existing background theories. Let's have a brief look.

2.3.3 Theory-Ladenness of Measurement

Suppose I want to determine the mass of my bowling ball. To that end, I might put it onto an ordinary balance and compare it with a known mass, such as a number of metal cubes each weighing 1, 10, or 100 grams. Things get more – much more – complicated if I want to know the mass of a distant star. Obviously, I can't put it onto any balance. So how do I measure it? The details are actually quite complicated, but we can roughly say that measuring the mass of a star involves various background theories. For example, if we want to measure the mass of a binary star, we first determine a center of mass between the two stars, then their distance from that center which we can then use, together with a value for the period (the time it takes to complete one orbit around each other) and a certain instance of Kepler's Third Law, to calculate the mass. In other words, in order to "measure" the star mass, we measure other quantities and use those values, together with certain equations, to calculate the mass. Obviously, the correctness of such a "measurement" not only depends on the correctness of other measurements (in this case, at least that of the period), but also on the correctness of certain background assumptions, such as Kepler's Laws. Measurement is not a simple and unmediated estimation of independently existing properties, but often a determination of certain magnitudes before the background of a number of accepted theories.

A moment's reflection shows that even in a simple measurement, such as that of temporal duration, background assumptions play a role. To measure the duration of a process A, we count the number of times that another process B is completed. Of B we assume that it exhibits strict periodicity, i.e., that its completion time does not vary from one to the next execution of the process. Familiar examples include the period from full moon to full moon. Today, the second as the unit of time is fixed by the number of oscillations of the cesium atom 133 (one second is about 9 billion such oscillations; see https://physics.nist.gov/cuu/Units/current.html). Since this behavior of cesium defines the unit of time, there is no way to verify by measurement that these oscillations really take one second. That they do is built into the definition of "one second."

One might think that the theory-ladenness of measurement raises methodological questions. Measurements are supposed to deliver data, which in turn can be used as evidence for or against some theory. Obviously, if the truth of a theory to be tested by data delivered by a certain measurement procedure is presupposed by that very measurement procedure, we face a circularity problem. Fortunately, this is hardly ever the case, and current consensus has it that the theory-ladenness of measurement poses no significant methodological problem. After all, since the assumption that certain background beliefs are true is unavoidable for any measurement procedure, we can't escape it. Moreover, as we will see in the next chapter, observation is also influenced by background beliefs.

Not only are background beliefs involved in the measurement process, there is an influence in the other direction as well: The measurement process can influence the measured. As a simple example, imagine a tiny amount of water the temperature of which is being measured by immersing a thermometer into it. Suppose the thermometer's housing is quite a bit colder than the water. If the quantities involved are just right, immersing the thermometer, and then waiting to see where the mercury column settles, will of course change the water's temperature, even if only by a small amount. Usually, the amount is so small as to be negligible. The situation is, however, different in the science of the very small – quantum mechanics (QM). There, we find the so-called measurement problem, which we will briefly discuss, together with other quantum oddities, in Chapter 11.

2.4 Conclusion

We saw in this chapter that neither naked observation nor measurement provide access to evidence that is fully independent from background beliefs. Seeing something *as* evidence typically involves bringing certain additional information to the table, as our example of the physician who can see lesions on the gum *as* Koplik spots, and thus *as* evidence for measles, shows. Such theory-ladenness of observation can perhaps explain certain forms of scientific disagreement, as we suggest in Chapter 4. Measurement exhibits a similar dependence on background theories, often of a highly sophisticated kind. In addition, the use of measurement raises questions about appropriate measurement scales and also also about the conditions under which measurement changes the values of the parameters we want to measure. We next turn to the question what to do with the evidence we have collected.

Note

1   According to the Standard Model of particle physics, the strong and weak force are responsible for the existence of atoms. See https://home.cern/science/physics/standard-model for a quick overview.

Annotated Bibliography

Peter Achinstein, 2001, *The Book of Evidence*. New York and Oxford: Oxford University Press. Arguing against traditional accounts of evidence, Achinstein introduces a distinction between four different notions of evidence and tries to show how this differentiated approach can solve various puzzles, such as the Raven Paradox and the Problem of Old Evidence.

Hasok Chang, 2004, *Inventing Temperature: Measurement and Scientific Progress*. New York and Oxford: Oxford University Press. The book contains several studies of the historical development of the concept of temperature between the seveenth and the nineteenth centuries. Among other things, the author argues that carrying out measurements not only leads to measurement results, but also to measurement conventions and determinations and refinements of scales.

Thomas Kelly, 2014, "Evidence," *Stanford Encyclopedia of Philosophy*. https://plato.stanford.edu/entries/evidence
      Discusses in greater detail the distinction between the phenomenal and the causal conceptions of evidence, which the author treats in a somewhat different way under the heading of evidence as a sign or mark of truth.

3

USES OF EVIDENCE

3.1 From Observation to Hypothesis

Empirical science often (but not always) starts with observations. You might notice that a squirrel searches through the trash bin in the park every day during your afternoon stroll. This makes you curious: Does the squirrel for some reason only look through the trash at that time, or does it forage at other times too? So you start observing the squirrel's behavior in a more systematic way: You spend long hours in the park, carefully cataloguing the squirrel's trash pursuits. Pretty soon, you see that your initial suspicion was correct – it only goes through the trash in the afternoons (you can rule out that the squirrel searches at night, because the trash is emptied every evening). You have found a regularity, and now you are curious about how to account for it. It quickly occurs to you that the squirrel probably only spends the effort of searching when there is a good chance to find food. Since the park is a favorite lunch spot for quite a few people, who often bring their dogs, some leftovers are bound to end up in the trash bin in the afternoon. By 1:00 p.m. (the time of your stroll), most people have left, the dogs are gone, and the area is fairly safe for the squirrel to commence its foraging. The regularity in the squirrel's behavior seems best explained in terms of the regularity exhibited in people's lunch behavior and that of their canine companions.

Admittedly, this little story is a far cry from what goes on in the empirical sciences, but it is close enough to extract some lessons from it. First, a scientist might notice something that deserves greater scrutiny. Often, in the case

This is Philosophy of Science: An Introduction, First Edition. Franz-Peter Griesmaier and Jeffrey A. Lockwood.

of scientific research, what we notice is influenced by background theories that we already accept (e.g., many animals exhibit activity patterns and squirrels seem quite adept at learning). Second, to confirm the suspicion that something unexpected is really going on, we resort to systematic observations (i.e., sitting in the park for hours on end). These will often, or at least sometimes, lead to the discovery of an interesting regularity, viz. squirrel foraging is concentrated in the afternoons. Third, the regularity prompts us to search for an explanation – unless we are dealing with extremely fundamental issues (more on that later), we assume that observed regularities aren't just brute facts. So we try to find an explanatory hypothesis (e.g., squirrels rummage in trash cans at times when food is abundant and predators are scarce).

But how should we go about finding appropriate candidates for explanations? This is a tricky question and perhaps not something for which there exists a general answer. Some people reason by analogy, others in terms of plausibility, etc. How a person comes up with a candidate explanation is a psychological question. There doesn't seem to be any foolproof methodology that guides the researcher from observed regularity to explanatory hypothesis. Rather, discoveries of explanations are often influenced by individual background beliefs, the person's creativity, and many other factors that defy a clear systematization. Thus, philosophers of science have routinely distinguished between the *context of discovery* and the *context of justification*. The context of discovery is characterized by the circumstances in which a scientist finds, or discovers, her explanatory theories. Because different scientists go about finding explanations in different ways, or so it has been argued, nothing of interest to the philosopher of science who is trying to find the universal norms that govern science, can be learned from how scientists discover their hypotheses.

However, the context of justification is a different matter. Once we have an explanation, in the form of a hypothesis or a more encompassing theory, we can legitimately ask whether or not we are justified in accepting that hypothesis or theory. On most accounts, a theory is rationally acceptable if it gets the data right for which it was invented, and also new data that hasn't been observed so far. In other words, we see what predictions the theory makes about the world and then check whether those predictions come to pass. If they do, good; if not, bad. How to spell out "good" and "bad" in this context is the subject of the next main section on theory confirmation.

Before we turn to those issues, we have to consider some complications in the role we assigned to observations so far, because especially in light of the earlier example, it makes it seem that what we decide to investigate is pretty unguided. In scientific practice, however, those things worthy of our attention are often suggested by already accepted theories. For example, in 1827, the Scottish botanist Robert Brown noticed that small particles suspended in a liquid were in constant, erratic motion, which could not be explained by any currents. This was noteworthy, because of the background assumption that things don't move without a cause. It was only in 1905 that Einstein explained the motion as the effect of molecules colliding with the particles. On the other hand, there are many facts not worth knowing about, as, for example, the number of blades of grass bent to the left by at most 33 degrees on Saturday afternoons on north-facing slopes in cities whose name has an "l" as their third letter when translated into English. And there is also the fact of the number of such blades of grass bent by at most 34 degrees, or 34.5 degrees, or on Saturday mornings, nights, etc. Clearly, there is an unbounded number of facts in the world, all of which are potentially interesting to someone. Given this unbounded number of facts, it is a good thing that our observational interests are guided and constrained by theories we already accept. Otherwise, we would be wasting a lot of time and other resources trying to explain facts that are not particularly important for understanding the world around us.

However, this selection of what is worth observing by referring to our already accepted stock of theories has a somewhat surprising consequence. If it is true that what we deem worthy of observing today is constrained by theories we accept, and if those theories were based on observations deemed worthy of making, which in turn were constrained by prior theories, and so on, all the back to the "beginning of science," it becomes quickly clear that our current theories are constrained by early observations through the historical trajectory of theories initiated by those observations. In other words, we have to acknowledge the possibility that what seems scientifically interesting today has been strongly shaped by the "starting point" of science. Had humanity been noticing things other than what it did notice, the scientific trajectory may have looked very different with the result that our current science might be different as well.

This thought may have consequences for the debate about scientific realism, which is the view that our best theories are at least approximately true. Even if this is so, it might be of little comfort for our belief that sciences provides us with a deep understanding of the world. We may well have true theories, but they may be true about facts that are as irrelevant to a real understanding of the world as are bent blades of grass.

3.2 Theory Appraisal

Suppose you found a theory which you consider to be pretty convincing. Others don't agree. What can you do? One thing to do is of course to show that the theory accounts for all the data – evidence – that are relevant to it. In particular, you are in good shape if you can show that your theory correctly predicts data that have not been collected yet. For example, suppose that a theory explains reductions in the Earth's temperature in terms of periods of volcanic activity during which emitted material reflects solar energy. Suppose further that in 1979, a scientist predicted that after an eruption of 0.5 to 1.0 cubic miles of ejecta, global temperatures would drop by 0.1 degrees Celsius. And then came the eruption of Mount St. Helens which spewed 0.7 cubic miles of material into the atmosphere – and the Earth cooled by 0.1 degrees. Well, the theory would be looking pretty good! How exactly does this process work – what is its logic, as it were?

3.2.1 Confirmation through Predictive Success

One influential proposal for how to understand this process is Carl Gustav Hempel's model of theory confirmation. It basically reduces confirmation to a two-step process. First, a scientist generates testable predictions from a theory. Second, the researcher determines whether or not those predictions are borne out by appropriate observations, which of course include measurement results and experimental outcomes. For Hempel, the first step was a matter of deductive logic. Here's an example.

Suppose that you want to defend the theory that the earth is spherical. The ancient philosopher Aristotle was one of the first to describe a test for this theory. The idea behind his attempt at confirmation is very simple. If the earth were spherical, we'd expect to see a ship that leaves the harbor to not only look smaller and smaller as it sails away, but to also disappear nonuniformly over the horizon. At the appropriate distance, we might be able to still make out the masts and the sails without seeing the hull of the ship anymore. So, the theory "The earth is spherical" predicts a partial disappearance of the ship. If you go to the harbor, you can actually make this observation – the ship disappears from the bottom-up, and the theory has been confirmed by this observation.

Hempel summarized and generalized this example, and others, into the following model of theory confirmation:

> Theory predicts Observation
> Observation is made
> Theory is (better) confirmed

The first thing to notice here is that confirmation is different from proof. If a theory entails an observation and the observation is made, it doesn't logically follow that the theory is true. To think otherwise would be to commit a logical fallacy. To see this, consider: "If it rains, the streets are wet. The streets are wet. Thus, it is raining." But the conclusion doesn't follow from the premises, because someone could have hosed down the streets, and that's why they are wet. However, if the prediction is borne out, we can say that the theory has been confirmed to a certain degree. This in turn means that we have some reason for believing the theory.

Here's another example following the same pattern. Suppose I want to confirm the hypothesis that all ravens are black. How would I do this? Following Hempel's advice, I infer a prediction from my hypothesis and then determine whether it is true or not. What follows from the claim that all ravens are black? One thing that comes to mind immediately is that the next raven I see will be black. Thus,

> Raven Theory predicts "Next raven is black"
> Observation "Next raven is black" is made
> Raven Theory is (better) confirmed

What's important about this second example is that the theory talks about all ravens – past, present, and future, and everywhere they exist. Suppose I have seen 200 ravens, and all of them have been black. Thus, I inductively infer my Black Raven theory. Now I want to confirm it by using Hempel's procedure, as we have just done. But, seriously, how much of a confirmation can that one black raven provide? Remember, my theory talks about all ravens.

3.2.2 Falsification to the Rescue

This problem led some philosophers to abandon any attempt to confirm theories and to look at the situation from a different perspective. The Austrian philosopher Karl Popper introduced the twin notions of *falsification* and *corroboration* to describe how we (should) go about selecting the theories we accept. The idea is fairly straightforward and starts with the observation that while we cannot conclusively confirm any theory, we can conclusively falsify a theory. Take our example involving the color of ravens. Perhaps 200 ravens might be good evidence for the claim that all ravens are black; 2000 black ravens would be even better evidence. However, no matter how many ravens you manage to observe, you can never conclusively establish that all of them are black. But you *can* conclusively establish that not all of them are black, should that be indeed the case. How? Simply find a raven that is not black. In other words, one counterexample to the claim that all ravens are black suffices to falsify it. The general model of this process is very similar to that of confirmation by prediction, with the important exemption that the predicted observation is not made:

> Raven Theory predicts "Next raven is black"
> Observation "Next raven is black" is not made
> Raven Theory is falsified

Popper made falsification the hallmark of his approach. Scientists, on his view, conjecture all sorts of theories and hypotheses. Whether any of them are worth keeping depends on two factors: How much do they tell us about reality, and whether or not they have been falsified. If we find a theory that has a lot of empirical content, and that we have not been able to falsify despite subjecting it to rigorous testing, we consider it as worth keeping. Popper's technical term for "not falsified after rigorous testing" is "corroborated." Thus, a theory is corroborated to the degree to which we have tried to falsify it and failed to do so. Highly corroborated theories are those worth keeping.

Why is having a lot of empirical content important, and how do we measure the content? First, theories with little to no content are not easily falsifiable. Suppose you restrict your raven theory in the following way: "All the ravens in my raven coop are black." Unless you have overlooked a raven in your coop, nothing can falsify this theory. It is so specific that there are not many falsifiers. As such, it is not a very interesting theory. So, let's be a little more daring: "All ravens in my yard are black." Ravens come and go, and so it might turn out that one morning, you get up and discover that a maroon raven is perched on a tree limb in your yard. However, even this daring theory is still a far cry from the original theory we considered, according to which *all* ravens are black. Clearly, it has a lot more content – talks about a lot more things – than the theory about the ravens in the coop. Since it talks about more things, there are more things that can falsify it. Thus, we can now answer, at least in principle, the question how we should measure the content of a theory. To a first approximation, the content simply is the set of things that could falsify it, were you to observe them.

To summarize. Theories that involve universal generalizations, such as "all ravens are black," can never be conclusively confirmed. They can, however, be conclusively falsified, or so Popper claimed. Theories that survive a lot of serious attempts to falsify them have proven their worth: they are corroborated. What science aims for is corroborated theories. Thus, we don't need a theory of confirmation. Problem solved.

3.2.3 Ravens and White Chalk

The falsification approach has a second advantage: It solves a notorious problem that arises for Hempel's model of confirmation, a problem of which Hempel himself was aware. The problem itself is a bit on the abstract side and will strike any working scientist as a "typical philosophical problem," but it is worth considering at this juncture because it is both clever and troubling. Imagine that an ornithologist had spent his career studying ravens around the world. He's seen a couple thousand of the birds, and every one has been black. So, he's working on a paper proposing that all ravens are black, but he's wondering what evidence he can use in addition to his observations of the birds. While lecturing in class, he looks at the chalk in his hand, stops speaking, and a big smile comes to his face. The ornithologist realizes that the white chalk supports his thesis. But how could this be?

Notice first that all ravens are black means the same as that if anything is a raven, then it is black. Thus, if it is not black, it is not a raven. In other words, all things that are not black are not ravens. Therefore, "all ravens are black" is logically equivalent to "all nonblack things are nonravens." Now, it seems clear that if a piece of evidence confirms a hypothesis H, that same piece of evidence should also confirm all those hypotheses which are logically equivalent to H. This widely accepted principle is known as *Hempel's equivalence condition.*[1] Suppose we call the hypothesis "All nonblack things are nonravens" H. H must be confirmed by any nonblack thing that is not a raven. For example, a piece of white chalk will confirm H. However, since H is equivalent to "All ravens are black," it follows that the piece of white chalk also confirms it! Let this sink in: The ornithologist's claim that all ravens are black is confirmed by the observation of a white piece of chalk. To many, this seemed strange enough to call it paradoxical, and so the problem received its name: Hempel's Raven Paradox.

Let us look at the paradox in explicit argument form:

1. Positive instances confirm a universal generalization, such as the generalization "All ravens are black."
2. Logically equivalent theories are confirmed by the same evidence.

3.   "All ravens are black" is logically equivalent to "All nonblack things are nonravens."
4.   A piece of white chalk is a positive instance of the generalization "All nonblack things are nonravens."
5.   Thus, from 1, it follows that a piece of white chalk confirms the generalization "All nonblack things are nonravens."
6.   Therefore, from 2, 3, and 5, it follows that a piece of white chalk confirms the generalization "All ravens are black."

The conclusion stated on line 6 gives rise to the paradox we are worried about. What happens if we replace the confirmation approach by the falsification approach? Somewhat surprisingly, the paradox doesn't arise. "All nonblack things are nonravens" is falsified by a nonblack thing that is a raven, which is the same as a raven that is not black. Of course, ravens that are not black also falsify the original hypothesis, "All ravens are black." Thus, the same piece of evidence falsifies the hypothesis in both of its logically equivalent formulations. There is not a whiff of paradox here. To see this more clearly, let's cast the falsification treatment of the "Raven Paradox" in explicit argument form as well:

1*. Negative instances falsify a universal generalization, such as the generalization "All ravens are black."
2*. Logically equivalent theories are falsified by the same evidence.
3*. "All ravens are black" is logically equivalent to "All nonblack things are nonravens."
4*. A nonblack raven is a negative instance of the generalization "All nonblack things are nonravens."
5*. Thus, from 1, it follows that a nonblack raven falsifies the generalization "All nonblack things are nonravens."
6*. Therefore, from 2, 3, and 5, it follows that a nonblack raven falsifies the generalization "All ravens are black."

Line 6* is not paradoxical at all! On the contrary, it is exactly what we would expect on the falsificationist approach. A nonblack raven *is* a negative instance of the generalization "All ravens are black." That it falsifies "All ravens are black" follows from 1*. Thus, in light of its ability to avoid the Raven Paradox, while keeping with the view that logically equivalent theories should be evaluated the same in light of the same evidence, the falsificationist approach seems preferable over the confirmation approach to theory acceptance. Again, for the falsificationist, we should accept a theory as long as it hasn't been falsified, despite rigorous testing. Tests on this view should only be used to rid science of false theories; they are useless for showing that a theory is likely to be true because a test revealed a positive instance.

It might be tempting to regard the falsificationist treatment of Hempel's Raven Paradox as a decisive victory for Popper's approach. It apparently can honor the idea that logically equivalent theories should be treated the same by the same evidence without leading to a paradox. Hempel's confirmation approach, in contrast, leads to a paradox. However, there are also serious problems besetting Popper's approach. To see this, let's return to the spherical earth example from earlier.

3.2.4 On Flat Earth and Bending Light

Consider a member of the flat earth society. He believes that the earth is flat. Obviously, he has to defend against the evidence from partially disappearing ships. For if the flat earth theory were true, we expect to observe just continuous shrinking, but not partial disappearance. However, we do observe partial disappearance. It seems that this observation falsifies the flat earth theory decisively. However, the flat earther has a move left, even if it will strike you as some sort of a parlor trick.

In response to the partially disappearing ship, he might say something along these lines: "Well, this observation is quite compatible with a flat earth. For example, if light didn't travel in straight lines, but was simply bent slightly toward the earth on its trip from the ship to the observer, then, after a certain distance, the light reflecting off the bottom of the ship would hit the water before it had a chance to reach the eyes of the observer. Thus, the bottom of the ship will disappear from sight before the masts will, even though the earth is flat."

What this guy is saying sounds outrageous – light being bent? (As we'll see, it is, as a general idea, not as outrageous as it may first seem, but it couldn't explain the ship's partial disappearance.) However, from the perspective of theory falsification, he has a point, namely this one: No hypothesis has observational consequences all by itself. There are always so-called *auxiliary hypotheses* that need to be in place as well. In our example, the "flat earth" hypothesis together with the claim that light travels in straight lines predict that a ship sailing away from the observer will shrink continuously and uniformly; it will not partially disappear. If the ship does partially disappear, I can blame one of the two claims: either that the earth is flat, or that light travels in straight lines. Either "falsity" would account for the partial disappearance.

You might still resist this move of blaming an auxiliary hypothesis (in our case, the claim that light travels in straight lines); it might strike you as cheating. But there are many other examples where this move is exactly the move to make. Suppose a group of physics majors gets a result in a lab exercise that contradicts some well-established theory. Clearly, we are not immediately going to overthrow the theory. Rather, we'll blame the students: They didn't set up the experiment correctly, they misread the measuring instrument, the instrument was broken, or what have you. The last thing we do is take them to have falsified classical mechanics! If we were to go there, all theories in physics, chemistry, and so on would continuously be falsified by legions of students doing lab exercises in colleges all over the US on a daily basis. Let's not go there.

The general lesson here is this. Theories and hypotheses always rely on auxiliary hypotheses in order to generate observable predictions. This phenomenon is known as *confirmation holism*. It was first explicitly discussed by the French physicist Pierre Duhem, who said in his book, *The Aim and Structure of Physical Theory*: "To seek to separate each of the hypotheses of theoretical physics from the other assumptions upon which this science rests, in order to subject it in isolation to the control of observation, is to pursue a chimera."[2] And while Duhem restricted his discussion to physics, it is clear that the underlying point generalizes: If we don't observe what the main hypothesis predicts, we can blame either the main hypothesis, or one or more of the auxiliary hypotheses. The outcome of the test does not tell us, however, which of the available hypotheses – main or any of the auxiliary ones – we should blame for observed discrepancies. In other words, hypotheses, and especially entire theories, which usually consist of many integrated hypotheses, cannot be conclusively falsified.

3.3 The Demarcation Problem

To be honest, we have simplified Popper's position to a considerable extent in order to introduce the notion of falsification. His actual view is much more sophisticated. To see this, consider a famous problem that motivated much of early twentieth-century work in the philosophy of science: the so-called *demarcation problem*, which is the problem of distinguishing empirical science from various forms of pseudoscience. For a group of philosophers known as *logical positivists* (a.k.a. logical *empiricists*), pseudoscience included metaphysics. The positivists sought to draw the distinction in terms of meaningfulness, which in turn was assessed by the verifiability of statements. For example, the statement, "The absolute is beautiful" sounds at first glance meaningful; after all, it is a grammatically well-formed sentence in English. But what exactly does it say? The positivists thought that if a statement has meaning, you should be able to determine whether it is true or false. If you can't, the statement is meaningless. Now, is it true that the absolute is beautiful? How would you verify it? If you think that this statement can't be verified, because you don't know, e.g., where to find the absolute to see whether it is indeed beautiful, then you should conclude, so the positivists contended, that it is meaningless. Of course, you might find the statement evocative, or it might resonate with you on an aesthetic level, and so have *some* sort of meaning. But what it lacks, according to the positivists, is *cognitive* meaning. Real science consists of statements that are cognitively meaningful, i.e., can be verified. This view has become known as the verifiability theory of cognitive meaningfulness.

Unfortunately, the verifiability criterion is too strong and rules out a lot of science. Consider again the claim that all ravens are black. We have seen earlier that to confirm this seems hopeless; verifying it is outright impossible, because you'd have to inspect all ravens past, present, and future. By the verifiability criterion for meaningfulness, the claim that all ravens are black is therefore meaningless. But that seems wrong. While "The absolute is beautiful" is rather obviously meaningless, "All ravens are black" is not. However, both are unverifiable. Thus, verifiability is the wrong criterion for distinguishing the meaningful from the meaningless.

Popper agreed with the positivists on the importance of distinguishing genuine science from pseudoscientific babble, but he didn't think that taking a detour through a criterion for meaningfulness was the right approach. So he opted for falsification instead. "All ravens are black" can be falsified, while "The absolute is beautiful" can't be. You can, in principle, find at a raven that's not black, but you can't find the absolute and see that it's ugly. Thus, the first statement belongs to science, while the second one doesn't, whether meaningful or not.

3.3.1 Progressive Modifications

However, we have also seen that due to the role of auxiliary hypotheses in theory testing, conclusive falsification is not possible either. A hypothesis can always be protected against falsification by denying one or more of the auxiliary hypotheses; in principle, one can even deny the observational statement with which the hypothesis has been found inconsistent. In the early 1800s, English chemist William Prout proposed that the atomic weights of the various elements are whole multiples of the atomic weight of hydrogen. It was well known though that some elements appear to have weights that are inconsistent with Prout's hypothesis. Chlorine, for example, was measured to have 35.5 times the weight of hydrogen. Prout remained undeterred and suggested that the chemical processes used to isolate elements were defective, and thus the chlorine sample was impure. In this case, assuming the truth of the main hypothesis was used to criticize and consequently modify the then prevailing experimental techniques that produced observational statements.

This raises the question of how one should decide what to modify in light of an inconsistency between theory and observational statements: the main hypothesis, one or more of the auxiliary hypotheses, or the observational statements which are based on the experimental techniques producing them? As an answer, Popper proposed that any modifications made should *increase the falsifiability* of the resulting theory. Since falsifiability is a measure of content – the more falsifiers a theory has, the greater its content – this amounts to the advice to modify in the direction of greater content, by either being broader in scope or more precise.

As an example, consider the theory C that all orbits of *celestial bodies* are *circular*. This theory is of broader scope than the theory P that all orbits of *planets* are circular, since planets are a particular kind of celestial body. Having broader scope, it has more falsifiers. On the other hand, it is also more precise than the theory E that all orbits of celestial bodies are *ellipses*, because circles are a kind of ellipses. Thus, C has more content than both P and E – it has more content than P by being more universal, and it has more content than E by being more precise.

Popper claimed that modifications that increase a theory's falsifiability contribute to the progress of science. The intuition behind this claim seems to be this: If we protect a theory from falsifications by modifying it in a non-falsifiable way, scientific development comes to a stand still. Yes, the current theory has been protected from falsification, but at the expense of stagnation. It's like going into the corner and pouting. Here's a sports analogy. Suppose you have been beaten at your favorite sport. Sure, you can look for excuses – the referee was unfair, there was too much wind, the ball was rigged, etc. And since these things seem to be always happening, you won't play anymore.

But if that's your reaction, you clearly won't make progress. Maybe the better strategy is trying to improve your game and then compete as hard and as often as you can, which of course increases your opportunities of being beaten. If you then stay unbeaten in ever tougher competition, it seems like you've made progress, certainly more than you could have made by frowning in the corner and not playing again. Of course, even if you stay unbeaten for a long time doesn't mean that you are a perfect player; there might still be better ones out there. But you have earned your right to play on for a while, to be kept on the team, as it were. And what goes for sports, Popper thinks, goes for theories. If we expose them to severe competition with other theories through tests, the ones that keep on winning earn their keep.

As mentioned earlier, Popper's technical term for staying unfalsified through severe tests is *being corroborated*. The corroboration of a theory does not provide a reason for believing it to be true, or even probable to any degree. Rather, it means that it has survived varied and severe tests, where severity is a function of the number of potential falsifiers for a theory. Assigning corroboration is simply saying that the theory is consistent with a set of statements that are currently accepted as basic. Thus, the degree of corroboration of a theory can change with changes in bodies of accepted basic statements. It is therefore clearly not a sort of truth value, because the truth of a statement is not in this way relative to what other statements are accepted.

3.3.2 Basic Statements

What statements are accepted as basic at a given time? In contrast to the positivists, who thought they have found epistemically privileged statements in so-called protocol sentences, which allegedly describe the experiential bedrock for all of our theorizing, Popper thought that what statements are accepted as basic depends on the experimental context and can change over time. He did provide a list of considerations that should be involved in any decision process about basic statements. First, they must be easy to test. In Prout's time, finding the atomic weights was thought to be easy. Second, they must be relevant to the theory undergoing testing. When Prout claimed that the weights of all elements are whole multiples of the weight of hydrogen, it was clear that a relevant basic statement is one about the weight of some element other than hydrogen. Third, we must keep in mind that their acceptance is provisional, and should the need arise, they too can be tested relative to other statements that are then accepted as basic for that purpose. The statement about chlorine was accepted as basic when it was used in an attempt to falsify Prout's claim. But the experimental methodology eventually became the subject of testing itself when it was determined that the available procedures for purifying substances were imperfect. Testing the standard purification procedures required other statements to be accepted as basic. Thus, on Popper's considered view, the process of falsification is a far cry from simply finding out that nature is in conflict with a hypothesis. Instead, it is embedded in a context of continuing criticism, where in principle nothing is exempt from criticism. There is no epistemic bedrock.

3.3.3 Moving and Burning

Popper's conception of empirical science as a process of continual criticism of bold empirical conjectures has been attractive to many working scientists. But is his model of how science grows through criticism borne out by historical evidence? Let's look at the Polish astronomer Nikolaus Copernicus' reaction to the problem of stellar parallax and the British chemist Joseph Priestley's introduction of negative weight.

Suppose you are in a driving car, approaching a city. You happen to look at two tall buildings far ahead on the right side of the road. One of them seems to be almost attached to the other, that's how close they look to you. However, once your car is driving right past them, you notice that the two buildings are actually dozens of yards apart; of course, as the car continues on its way and you look back, the buildings seem to be closer together again. Something very similar should happen, if heliocentrism is true. As the earth moves around the sun, stars that look to be very close to each other at one point during the year should look to be much farther apart at a later point, at a time when the earth is moving "right past them." This apparent change in the distance between the stars, as observed from the moving earth, is the phenomenon called "stellar parallax." The problem was, for Copernicus and his contemporaries, that no stellar parallax had ever been observed! Thus, the opponents of heliocentrism argued that the earth can't be moving through space and around the sun. To them, the absence of an observable stellar parallax clearly falsified heliocentrism.

In response, Copernicus resorted to the move we discussed earlier: Blame some auxiliary hypothesis! He simply proposed that the absence of an observable stellar parallax is due to the fact that the stars are much further away from us than we thought. If so, then the distance travelled by the earth would be negligible in terms of observing parallax. During his time, most estimates of the size of the universe put the stars, which were thought to be located at the outer edge of the universe, at a distance of about six times that between the earth and the moon (they were off by about one quintillion-fold). In such a small universe, one would expect to observe a stellar parallax. In effect, Copernicus pointed out that to infer that we should observe stellar parallax, two claims need to be true: that the earth moves around the sun *and* that the universe is sufficiently small. Thus, he was able to blame the assumptions about the size of the universe for the failure to observe stellar parallax. Heliocentrism had not been falsified.

There is another infamous case in which an important scientist tried to save a theory from counterexample, but in which the attempted "rescue mission" led to a rather curious claim. We are thinking of Joseph Priestley, a lifelong defender of the so-called *phlogiston theory* of combustion. According to this theory, when things burn, they release a very subtle substance, called "phlogiston." The more phlogiston a material contains, the easier it burns. Given the importance of combustion for chemical experiments, phlogiston was a central concept in early chemistry. In the 1780s, the French chemist Antoine-Laurent de Lavoisier conducted experiments in which he burned mercury. Upon measuring the weight of the residue, he realized that it was heavier than the sample of mercury he started with. This was extremely puzzling: If combustion involves the release of a substance, why should the resulting "ashes" be heavier than the material before it was burned? This fact seemed to conclusively refute the phlogiston theory.

Priestley, however, had an "ingenious" idea to save the theory. Phlogiston, so he claimed, has negative weight. Thus, since phlogiston is released during combustion, the residue will be heavier than the original sample. Of course, since most things that we ordinarily burn, such as wood, also contain water, their ashes will be, all things considered, lighter than what we put into the fireplace, because the water evaporates. Obviously, Priestley made the same move as Copernicus. In order to observe weight loss during combustion, when adjusted for the loss of other components, two things have to be true: phlogiston is released during combustion *and* phlogiston has positive weight. Thus, he was able to blame the assumption about the weight of phlogiston for the failure to observe weight loss. The phlogiston theory had not been falsified.

3.3.4 Lucky Modifications

We quite deliberately phrased the ending of Copernicus' successful rescue of heliocentrism and the ending of Priestley's attempted rescue of the phlogiston theory almost identically, only replacing some of the words, in order to bring out how eerily similar the two episodes are. And yet, from today's perspective, Copernicus' rescue appears heroic, while Priestley's seems foolish. For one, we now know that Copernicus was right, and that Priestley was wrong. However, we need to be careful not to let hindsight play a role in evaluating their rescue attempts. From a logical perspective, they seem identical. Given their respective historical contexts, both the claim that the universe is vastly bigger than usually estimated, and that phlogiston has negative, rather than the standardly assumed positive, weight, are hopelessly *ad hoc*. It's just that Copernicus got lucky and was proven right by subsequent developments, while Priestley had massively bad luck and now looks like a fool.

Could their respective fates have been predicted at the time? If Popper is right about legitimate vs. illegitimate modifications in the face of observational counterevidence, it would have to be the case that Copernicus' modification was arguably progressive, while Priestley's was not. Remember that a modification is progressive to the extent to which the modified theory has more content, and thus faces more potential falsifiers, than its predecessor. Copernicus' modification postulated a very large universe. Priestley's modification postulated a substance with negative weight. Regarding Copernicus, it is difficult to determine whether heliocentrism in a large universe has more falsifiers than heliocentrism in a small universe. Thus, it is difficult to say whether this modification was indeed progressive in Popper's sense. The difficulty might reside in Popper's choice of the "scientific unit" that is to be judged as progressive or not. The unit for Popper is an individual theory. His student, Hungarian-born Imre Lakatos, proposed using a wider unit, viz., a *research programme*. Perhaps Copernicus' modification can be seen as progressive in virtue of being embedded in a progressive research programme, as we'll discuss in Chapter 15 on scientific progress.

Let's now turn to Priestley's negative weight. There is one sense in which this proposed modification is not progressive, but rather badly ad hoc. Phlogiston was supposed to be the only substance that has negative weight. But could this hypothesis at least be falsified? In Copernicus' case, the development of increasingly powerful telescopes provided good empirical reasons for adjusting the estimates about the size of the universe upward, until, by current estimates, we arrived at 91 billion light years. In principle, it could also have been falsified. This would seem to be impossible in the case of phlogiston. The obvious way would be to isolate phlogiston and then try to weigh it – but with what? We don't have instruments for determining the value of negative weight. Moreover, with negative weight, would phlogiston also have to have negative mass? (Remember, weight is simply a function of mass and the gravitational constant). What would that be? Sure, the negative weight idea saved the theory from the mercury counterexample. But it would have been quite obvious that independent evidence was elusive if not impossible. In light of this, perhaps the right verdict is to say that the statement "Phlogiston has negative weight" looks quite a bit like the statement "The absolute is beautiful." Neither one can in any clear way be falsified. Thus, by Popper's criterion for progressive modifications, Priestley's modification fails, as it introduces a nonfalsifiable, and thus merely protective, hypothesis.

3.4 Conclusion

This completes our discussion of the use of evidence. We discussed the role of evidence in theory generation and seen that much of this role there depends on a scientist's psychology and other factors. Theory confirmation, on the other hand, is supposed to be independent of such psychological factors. However, there are certain logical complications that afflict the relation between theory and evidence (e.g., the raven paradox). Popper's falsificationism has promise to help with some of those problems. Whether his picture of science as a process of critical inquiry that discards theories one by one can be generally accepted is doubtful. The question will be revisited in Chapter 15. But for now we delve further into evidence to consider the ways in which it can be evaluated and why sharing it with fellow scientists doesn't necessarily resolve disagreements among them.

Notes

1 Richard Swinburne, "The Paradoxes of Confirmation – A Survey," American Philosophical Quarterly, 1971, Vol. 8, 318–30.
2 French original published in 1914; English translation in 1954 by Princeton University Press. Quote from p. 199 f.

Annotated Bibliography

Vincenzo Crupi, 2020, "Confirmation," *The Stanford Encyclopedia pf Philosophy*. Available at https://plato.stanford.edu/entries/confirmation A detailed presentation of many of the technical problems surrounding the notion of confirmation. It also includes many more details on Hempel's model.

Brandon Fitelson and James Hawthorne, 2010, "How Bayesian Confirmation Theory Handles the Paradox of the Ravens," in E. Eells and James H. Fetzer (eds.), The Place of Probability in Science, Boston Studies in the Philosophy of Science 284. Available at http://fitelson.org/ravens.pdf.
  A thorough discussion of the raven paradox from the perspective of Bayesian confirmation theory, arguing that emphasizing the important difference between no confirmation at all and a small amount of confirmation resolves the paradox.
  Carl Gustav Hempel, 1945, "Studies in the Theory of Confirmation," *Mind* 54(213): 1–26 and 54(214): 97–121. In this groundbreaking paper, Hempel develops his model of confirmation and introduces the Raven Paradox.
  Karl Popper, 1934/1959, *The Logic of Scientific Discovery*. London and New York: Routledge 2002. The classic statement of falsificationism, this is one of the most influential books in the philosophy of science. Popper argues for a decisive break with attempts to develop a model of confirmation, replacing it with a process involving bold conjectures and severe testing.

Willard van Orman Quine, 1951, "Two Dogmas of Empiricism," *Philosophical Review* 60: 20–43.
  Among other important contributions, this seminal essay introduces the idea of confirmation holism.

4

EVIDENCE, RATIONALITY, AND DISAGREEMENT

4.1 From Weak to Strong Evidence

It is widely accepted that there is good and bad evidence. In other words, we recognize a quality spectrum of evidence from the pretty weak to the exceedingly strong. At the weak end, we find what is often called "anecdotal evidence," while at the strong end, we see the results of double-blind clinical trials. To understand the quality differences as we move through this spectrum, we need to pay attention to the various types of error possibilities that any evidence might leave open. Such attention to error is important because of the function we assigned to evidence in Chapter 2. There, we claimed that evidence can indicate the existence or presence of various things, processes, properties, and facts. For example, tracks in the snow can indicate the prior presence of a mule deer, and/or the speed of its movement (the spacing of the tracks – the gait – allows us to infer its speed). We use evidence in such a context precisely because many other "methods" of trying to find huntable deer turn out to be riddled with error possibilities – we have in mind here such methods as reading tea-leaves or using a divining rod. Now it is obvious that using evidence instead of these "methods" is only going to be useful if the error possibilities are reduced by doing so – if, for example, one is less likely to go off in the wrong direction by following deer tracks than by consulting the divining rod.

Intuitively, following the tracks is clearly less prone to error than the other "methods." However, some empirical evidence can also be misleading.

This is Philosophy of Science: An Introduction, First Edition. Franz-Peter Griesmaier and Jeffrey A. Lockwood.
© 2022 John Wiley & Sons, Inc. Published 2022 by John Wiley & Sons, Inc.

Instances of misleading evidence are perhaps concentrated at the weak end of the quality spectrum, mostly associated with anecdotal evidence, but they appear in other contexts as well. The most prominent errors are related to the presence of so-called *confounding factors* and of *biases* of various sorts. Briefly, a confounding factor is a feature of the world which influences what the evidence indicates without being evidence itself. Thus, we may take the evidence to indicate A, when in actuality, whether or not A is the case depends on further factors. Modern cars, for example, are standardly equipped with a mileage indicator – a read-out that tells the driver how many more miles she can go before running out of gas. Those indicators reflect the gas usage of the recent past together with the amount of gas left in the tank. However, whether or not the car can really go for another 50 miles depends on whether or not the past driving conditions – absence of strong headwinds or a flat landscape, for example – continue to be operative. Put differently, indicating a remaining distance of 50 miles not only reflects the remaining amount of gas and recent gas mileage, but also the absence of strong headwinds and of steep climbs (these are confounding factors).

Evidence can also reflect biases. In particular, how evidence is interpreted – what it is taken to show or how it is weighted – can be influenced by somebody's background beliefs, either implicitly or explicitly. This may come as a surprise at first – after all, evidence is evidence. But we have already encountered in the last two chapters several reasons for why such a simple view of evidence is mistaken. First, the same data can be seen as indicating different things, depending on the background knowledge of the observer (recall the Koplik spots). Second, one might have a special interest in the truth of a particular theory (e.g., one might have developed that theory and achieved fame) and thus take seriously the data that constitute evidence in its favor while downplaying the "bad" data, a phenomenon called *confirmation bias*. This exact concern was raised 400 years ago by the English philosopher Francis Bacon who wrote in *Novum Organum*: "The human understanding when it has once adopted an opinion (either as being the received opinion or as being agreeable to itself) draws all things else to support and agree with it."

Indeed, Bacon was one of the first intellectuals to clearly explicate the problems of bias. As a crucial voice in the rise of empiricism, he admonished that we must clear our muddled minds to make objective observations. Bacon argued that four "idols of the mind" warped our perceptions and interpretations:

- *Idols of the tribe* by which Bacon meant the tendency to mingle our species' inherently distorted perceptions with the true nature of the world.
- *Idols of the cave* which were those ideas acquired through social conditioning and individual experience. Bacon was alluding to Plato's allegory of the cave in which we see only flickering shadows of reality.
- *Idols of the marketplace* which comprised the ways in which language twisted our perceptions: "the ill and unfit choice of words ... plainly force and overrule the understanding, and throw all into confusion."
- *Idols of the theater* which referred to people uncritically repeating the scripts of conventional wisdom. Shakespeare wrote: "All the world's a stage/And all the men and women merely players" which captured Bacon's contemporaneous contention that people lived in fictional worlds unchallenged by first-hand experience.

The discussion in the following sections will bring us into the present age and identify various confounds (short for confounding factors) and biases associated with the kind of evidence under consideration. It will become clear that the principal reason for moving from anecdotal evidence to more sophisticated forms is grounded in the desire to develop increasingly more stringent safeguards against the distorting effects of confounds and biases.

4.1.1 Anecdotes

Let's begin with the simplest and weakest version of evidence: the anecdote. The word originally meant "secret" or "private" stories, which still captures the essence of the idea. Anecdotes are stories about events that someone witnessed. There are a number of problems with drawing inferences from anecdotes, many of which revolve around what might be called *"the person who"* fallacy. That is, anecdotes often take the form of: "I know a person who..." or "I heard about a place where..." or "I read about a time when..." Consider the following anecdote:

> Erin was allergic to cats until we took her on a family trip to Egypt, where she briefly contacted ancient progenitors of house cats at a hotel. When she got home, this fleeting dose of ancestral cats had cured her of the allergy. So, we are developing a homeopathic remedy for cat allergies so others can benefit from our experience.

Why would Erin's parents try to develop this kind of treatment? Assuming that they are benevolent, their reasoning might be as follows: "We know a person, our daughter, whose cat allergy subsided after she interacted with ancestral cats while in Egypt. Thus, this exposure to ancestral cats must have cured her allergy."

The first problem with this inference is that it is an instance of the fallacy called *post hoc, ergo propter hoc* (or *after this, therefore because of this*). That is to say, it can be an error to conclude that because X happened before Y occurred, Y was a result of X. Lots of events preceded the relief of Erin's allergies, including a transoceanic plane trip, a dinner of lamb stew, and countless other happenings, any of which could be credited with the cure – and none of which were likely to have had any effect. Her parents might want to assert that they had a temporal control with regard to the disappearance of their daughter's allergy, but given the number of *confounding* events between her last allergic reaction and the apparent cure, the contact with Egyptian cats cannot reasonably be claimed to be the likely cause of the subsiding of the allergy.

Second, anecdotes often refer to a single subject, such as Erin. And drawing any general conclusions from a single individual or event is an absurdly *hasty generalization*. For example, just because you got lucky and scored a 100 on your last exam without having studied much, you ought not infer that it's better not to study for exams. One event cannot serve as supporting evidence for a general conclusion even about events of the same kind. Consider also that we generally hear nothing about negative results. What about the hundreds of people with cat allergies who travelled to Egypt, encountered ancestral cats, and returned home unchanged? Or what about the countless students who didn't study and failed their exams?

Third, various cognitive biases are not easily avoided when relying on anecdotal evidence. One such bias is the so-called *availability heuristic*: When we evaluate the efficacy of a cure or the causal effects of some other event, we often rely on information that is easily available to us, which in turn often involves unusual occurrences. For example, smokers in defending their habit might refer to the fact that the comedian George Burns smoked a dozen cigars a day for 70 years without developing lung cancer, or that the actress Dana Reeve died of lung cancer but never smoked. These cases are outliers and do virtually nothing to undermine the causal connection between smoking and lung cancer. Cognitive scientists have suggested that our recollection of some fact (e.g., George Burns's cigar consumption) fallaciously leads us to believe that this must be important – otherwise, why would we have remembered it? We are particularly susceptible to this mistake when an event is vivid and emotional, as might be the case for parents whose daughter seemed so happy when she encountered the Egyptian cats without sneezing and itching.

And finally, in our allergy case, you didn't travel to Egypt nor did you see Erin's purported cure. Rather, you're relying on second-hand information, the veracity of which depends on the reliability of the person providing testimony. In legalistic language, you're counting on hearsay rather than you being an eyewitness to the events. Moreover, imagine that her parents are influenced by the confirmation bias and that their impassioned account makes you prone to the problem of the availability heuristic. It becomes quite clear that the anecdote is a very weak reason to draw the conclusion of an association between ancestral cats and allergies to felines.

Summing up, it is abundantly clear why anecdotal evidence is the least useful form of evidence. Neither confounds nor biases can be easily eliminated, and the risk of drawing erroneous conclusions is very high. In addition, anecdotes introduce testimony, which by itself raises difficult epistemological questions. For example, when using testimony as a source of evidence, one must be in a position to assess the reliability and trustworthiness of the witness.

4.1.2 Observational Studies

This last concern – the matter of reliable testimony – gives rise to two stronger forms of anecdote that the scientific community generally accepts: natural histories and case studies. In both of these, there is no purposeful intervention and no control in the experimental sense of leaving some subjects untreated.

4.1.3 Natural History

Natural history is an important and accepted practice in the life sciences (as well as geology, and astronomy). It took off during the late seventeenth century with the work of the Italian Gianluigi Buffon, when the term "history" was synonymous with "description." He described in great detail the primary quadrupeds and birds, as well as facts about chemistry, mineralogy, and historical cosmology. Even today, natural history involves the careful observation of the natural world, often with a systematic approach so that the features are reported as objectively and clearly as possible. Unlike the anecdote, natural history is conducted by an individual with particular expertise in a field and with experience in making less biased observations, and the report contains considerable detail to provide an in-depth account of the context in which the phenomenon occurs. Also unlike anecdotes, natural histories usually pertain to *representative* instances. Rather than seeking exceptions to a well-documented body of observations, the natural historian is looking for typical cases of a poorly documented class of phenomena. From the accumulation of such data, the scientist may then infer patterns and relationships that sometimes provide the raw material for designed experiments.

Perhaps the most famous natural historian was Charles Darwin, who wrote in a letter to a colleague, Joseph Dalton Hooker, in 1844:

> Besides a general interest about the Southern lands, I have been now ever since my return engaged in a very presumptuous work & which I know no one individual who w[ould] not say a very foolish one. – I was so struck with distribution of Galapagos organisms &c &c & with the character of the American fossil mammifers, &c &c that I determined to collect blindly every sort of fact, which c[ould] bear any way on what are species. – I have read heaps of agricultural & horticultural books, & have never ceased collecting facts – At last gleams of light have come, & I am almost convinced (quite contrary to opinion I started with) that species are not (it is like confessing a murder) immutable.
>
> (https://cudl.lib.cam.ac.uk/view/MS-DAR-00114-00003/1
> accessed 8/28/2018)

Darwin gathered observations of the natural world to assemble a collection of facts about, among other things, morphological differences and similarities among various animals, from which – along with the very basic notions of heritability, variation, and selection (much of which he derived from the "natural" histories of breeding domestic animals) – he constructed the conceptual framework of evolution. Today, the sequencing of genomes leads to a flurry of molecular natural history, which provides the raw material for developing phylogenetic patterns and explanations of the evolutionary history of various lineages.

In what sense do natural histories provide better evidence than anecdotes? Remember the major pitfalls of using anecdotes – uncontrolled confounds, rampant biases, and hearsay. So, are natural histories better able to correct for those problems? First, anecdotes are often used as evidence for causal claims, such as in the case of Erin's cat allergy. It is especially in the context of trying to support causal claims that confounds pose a problem. Natural histories by themselves hardly ever focus on establishing causal claims – they are most often just descriptions of nature. As such, the problem of confounds doesn't really arise. Second, there remains the risk of bias, particularly with respect to what a researcher pays attention to. To an extent, relying on reports from trained observers reduces the likelihood of bias. In addition, the precise descriptions of both the object and circumstances of discovery allow, at least in principle, for others to check the veracity of the reports collected in a natural history. Finally, there is the complicated issue of testimony. Historically, the value of a piece of testimonial evidence was often tied to the social status and character of the witness. Today, we hope to replace this appeal to social status with something more relevant, such as expertise. Expertise in turn is bound to institutional features, such as possessing relevant credentials. This gathering of natural observations by field scientists has a parallel practice in other realms of research – the case study.

4.1.4 Case Studies

Typically, a case study consists in the detailed examination of an individual system, such as a human being with a particular disease (especially if the disease is rare), and the subsequent attempt to draw some general lessons about cases of a similar kind. This latter process of inference is often called *extrapolation*, which is an inductive inference. This approach is most commonly employed in the social and health sciences (anthropology, clinical psychology, education, medicine, political science, and sociology) where a researcher's capacity to shape circumstances may be very limited, unlike in many areas of the biophysical sciences, which often rely on designed experiments. However, even in physics, we find some case studies. Cosmology is a case study, as there is only one universe (as far as we know).

Consider a science like ecology. Ecological units, which include individuals, populations, ecosystems, landscapes, and many more, might be different with respect to the number and importance of variants they admit. There is, for example, just one Yellowstone National Park, so that studying it constitutes a case study, the results of which cannot easily be extrapolated. On the other hand, mountain streams might exhibit important similarities across different locations that allow the extrapolation of information gleaned from studying one (or a few) of them.

A case study should provide grounds for extrapolating what has been learned from one case to other, similar cases. This is where things can get tricky. We already identified the problem of *hasty generalization* in the context of discussing anecdotal evidence. There, we said that it is inadvisable to draw general inferences from just one individual or event. But isn't that exactly what we do when using case studies as evidence? Well, not quite, or at least not always. To see what's going on, we need to separate two different kinds of inferences.

First, we might infer that the *regularities* found in one case study extrapolate to other cases of a similar kind. This practice is justifiable to the extent to which we have reason to believe that the class of cases over which we extrapolate does not vary along the relevant dimension(s). For example, if you were interested in how islands become colonized by plants, you might want to know whether or not pea-sized seeds of various trees float on salt water. In this context, you do not have to worry about possible variations in color across different seeds (as long as the color is not, for example, correlated with dryness of a seed which could change its density). Variations in color are irrelevant to floating behavior. On the other hand, if you wanted to know whether seeds could be transported through adhering to different substrates, variations in material are relevant. You can't extrapolate from the observation that a piece of balsam wood floats to the conclusion that a similarly sized hunk of lava could carry a seed (unless it was actually pumice). The justification of your extrapolation depends on the uniformity across cases along relevant dimensions.

Second, you might want to extrapolate some *causal connections* you have found in one case to other cases. For example, suppose a medical case study suggests that a patient developed a certain form of lung disease because he was exposed to toxic fumes at his welding shop. Extrapolating, you infer that other people working in similar welding shops are at risk of developing the same disease. This inference seems at first glance to be unqualifiedly justified. After all, welding processes using the same equipment can reasonably be expected to produce similar fumes. However, even if we suppose that the regularity (welding in shop → development of disease) extrapolates, the causal claim might not. For example, if you notice the same regularity but an absence of lung disease in another shop, you will weaken the original causal claim (that the disease was caused by the fumes) and realize that there must have been confounding factors. Perhaps the first shop was situated over a geological formation the emitted lots of radon, for example. In other words, a causal extrapolation from a case study is highly defeasible. This is not bad news, though, for this possibility allows us to test causal claims by looking at what seem to be similar cases and then try to discover the relevant confounds and thus, by a process of elimination, the real culprit. In a nutshell, similar cases can serve as replicates, through which we approach evidential standards that are similar to those involved in experimental evidence.

In general, it would seem that there is an inverse relationship between the degree to which extrapolation from one case to another is legitimate and the number and kind of relevant variations between the cases. Extrapolating from Yellowstone National Park to other landscapes might be fraught with risks of drawing erroneous conclusions, while extrapolating from one mountain stream to others within the same range of altitudes (say, between 5000 and 9000 ft), latitudes (40 to 60 degrees North, rather than equatorial), and similar climates (say, coastal mountain ranges, rather than midcontinental) might be less risky, especially if no causal claims are involved. In a graph (Figure 4.1), the simplest relation between variation and extrapolation may be specified as follows:

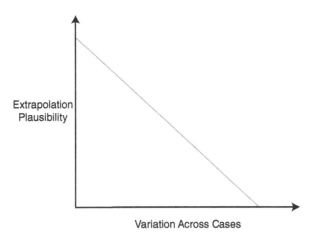

Figure 4.1 Extrapolation risk.

Of course, this is extremely simplified. The graph omits the role of the different dimensions of variation, some of which are more relevant than others to the feature we intend to extrapolate using a case study. Moreover, it treats all features of a case similarly, while in reality, some features might be more relevant than others. As an extreme example, think about a medical case and features pertaining to the duration of exposure to a possible pathogen versus the zodiac sign of the person so exposed (a less extreme example would be the duration exposure versus the age of the individual). It would take us too far afield to work out the details of the various tradeoffs between degrees of variation, the relevance of a dimension of variation, and the centrality of features of a case. Suffice it to say that different judgments about the best tradeoffs are a plausible source of disagreement among scientists that use the same case studies as evidence for claims about general regularities or about causal connections.

In terms of the three risks of confounds, biases, and hearsay, so prevalent in anecdotes, case studies fare about the same as natural history. Although these risks persist, they are markedly reduced by the involvement of proficient investigators.

Case studies can be particularly useful under the standards of Popperian falsification in which a single observation contrary to a proposition is sometimes deemed sufficient to invalidate the claim. Suppose there is widespread agreement that a newly emerged disease is 100% deadly (such as Ebola in its early days). A single case of a patient recovering would both falsify the absolute lethality of the disease and stimulate further research into what made this individual different.

We have seen that different kinds of evidence, from anecdotes to case studies, have different chances of mitigating the risk of error. If we also think that it is necessary for a rational response to evidence that we avoid error as much as possible, it becomes clear that case studies and natural histories should be afforded much more evidential weight than mere anecdotes. However, their capacity for controlling confounds and bias is still not as pronounced as it is in designed experiments. How this works is the topic of Chapters 6 and 7. In the remainder of the current chapter, we will address some general features of the relation between evidence and rationality, and in particular, how different responses to the same evidence can be more or less equally rational.

4.2 Evidence and Rationality

In Chapter 2, we developed a nice and clean picture of how evidence lets us learn something about the world. Evidence is, due to causal relations among different features of the world, a good indicator about those parts of the world that we do not, or cannot, directly observe. Unfortunately, this picture is a bit too nice, because it is too simple. The additional connection between evidence and rationality, which we introduced through our discussion of how different kinds of evidence give better or worse reasons for believing something, reveals that sometimes, acceptable evidence may not be always hooked up in the right way to the features we want to know about. Here is the problem.

Suppose you are a detective trying to solve a gruesome murder. You secure the bloody knife with which the victim was stabbed to death in his house and hand it over to forensics. You can't believe your luck: DNA found on the knife could be matched to a known felon! You have, in your hands, clear evidence that Mack did it. Clearly, you are rational in believing that Mack was the perp and that you should send out your cops to arrest him. However, when they get to his address, his wife tells them that Mack's not there anymore. He died, a couple of weeks ago, from a stroke. So he couldn't have done it! Something must have gone wrong in the lab, or Mack was somehow framed, or the actual perp has DNA indistinguishable from Mack's (he might've had an identical twin who nobody knew about), or some combination of factors. Thus, your belief that Mack did it turns out false, despite the seemingly good evidence you had. Does this mean that you were irrational in believing that Mack did it on the basis of the original forensic evidence?

Most people think the answer is no. As far as you are concerned, you have correctly proportioned your belief to the evidence. It's just that your evidence was bad. More precisely, it was *misleading evidence*. However, if you believe something on the basis of misleading evidence, and you don't know that it's misleading, you can still be rational. Therefore, one can be rational in believing a hypothesis on the basis of evidence that is misleading, or, as we can also say, evidence that is not veridical, i.e., not hooked up with the relevant facts.

When we consider the history of science, we quickly realize that forming hypotheses on the basis of misleading evidence is not the exception, but looks more like the rule. For example, until 1543, the year in which Copernicus published his new heliocentric astronomy, many believed that the Earth was stationary and at the center of the universe. There was good evidence for that belief: It surely seems as if everything is moving around us, and we surely don't feel as if we were on a merry-go-round. There is no sensation of being on an object which rotates (at the equator) at about 1,000 miles per hour and shoots through space at about 67,000 miles per hour! But the evidence is misleading. Observational evidence also strongly suggests that solid objects consist of continuous matter with no gaps. But if contemporary physics is right, this is a false impression again. Solid objects consist mostly of empty space – for example, the distance between electrons and the nucleus in an atom is huge compared to the size of either. At any rate, the history of science is (largely) the history of evidence misleading us into accepting false theories. The detective from above is in great company.

What we should conclude from these observations is not entirely uncontroversial. On the one hand, we do think of evidence as an objective indicator of unobserved features of the world. But clearly, sometimes evidence is not actually such an indicator, but only believed to be one. This suggests of course that being evidence is not simply a matter of standing in the right causal relations to what it is evidence for. At the very least, it has to be believed to stand in such a relation – before Mack's fingerprints were discovered, they were not evidence for the claim that he did it. And sometimes, beliefs about evidence are simply false. For those reasons, some have argued that evidence really is just that which gives a human observer, be it a detective or a scientist, *reasons for believing* that something is the way in which the (alleged) evidence indicates it to be – Mack's the murderer, the Earth's at rest, and objects are entirely solid. On such a view, evidence is defined by its capacity to *provide justification* for our beliefs, and not by its actual causal connections to features of the world. This view, which defines evidence in terms of its power to justify our beliefs, even if those beliefs are false, is sometimes called the *phenomenal conception of evidence*. It is in tension with the other view of evidence, which we considered in Chapter 2, according to which being evidence is a matter of being causally connected to features of the world in the right ways. We can call this the *causal conception of evidence*.

The causal conception has the advantage that it makes sense of our use of evidence as a publicly available, neutral arbiter when theories, or beliefs in general, come into conflict. The phenomenal conception has a disadvantage on this score. For if evidence is what we simply believe to be connected with the facts in the right way, then evidence might be neither neutral nor publicly available. First, if my beliefs about what constitutes evidence are influenced by my background beliefs, and if you have different background beliefs from mine, we might not even agree on what the evidence is! What counts as evidence for me might not count as evidence for you, and in such a case, evidence is not neutral. Second, if we include how things seem to me as being part of my evidence, then not all evidence is publicly available, for you cannot observe how things seem to me. We will revisit these problems later. For now, let us return to the connection between evidence and rationality and look at some of the details.

There is an influential view of epistemic rationality, called *evidentialism*. In a nutshell, the view holds that a person is rational in what she believes to the extent to which she respects her evidence. Remember Bob, the climate scientist, from earlier in Chapter 1. Despite the evidence against his own theory, he doesn't give it up. He is thus irrational. Evidentialists would account for this by saying that if two people have the same evidence, then they are equally justified in believing that which the evidence supports. In other words, epistemic rationality is simply a function of the evidence. Fix the evidence a person has, and you have fixed what it is rational for the person to believe.

Thus, if two rational people have the same evidence, they *should agree* on what to believe, including what theories to accept. However, we know that this doesn't happen in all cases. There are many genuine scientific disputes, and often they involve scientists who have the same evidence! How could this be? Let's consider at least two answers to this vital question for science, the second of which can be developed in two different ways:[1]

(i)   The simplest answer is that one of the scientists involved in such disputes, or maybe all of them, are irrational – they must fail to base their beliefs in the right way on their evidence. Many think that this answer is unsatisfactory, because on the face of it, the people involved seem to be able to rationally defend their views. At the very least, if we focus on the same evidence, the different sides in a dispute often seem equally reasonable. So we need another answer.

(ii) Maybe there is a difference between the bodies of evidence that the scientists use after all. In other words, while it may be true that the evidence explicitly referred to in a dispute is ostensibly the same for all the debating scientists, there is "something" in the background that makes a difference to how the shared evidence gets used in deciding which theory to accept. But what could such a "something" be? Roughly, there are two broad possibilities: (a) the parties in disagreement have different background beliefs, which somehow influence their evidential base, or (b) they assign different, so-called *prior*, probabilities to the theories under dispute. In the next couple of sections, we will try to understand the details involved in possibilities (a) and (b), and the consequences flowing from them for the question of how evidence bears on theory acceptance.

4.3 Explaining Scientific Disagreement

As intuitively plausible the slogan "follow the evidence" initially seems, our last considerations strongly suggest that the slogan is too vague. The existence of scientific disputes seems to show that different people can apparently "follow the evidence" equally well and yet end up believing different things. Our next task is to find out exactly how this can happen. We already distinguished two broad possibilities. First, it might be the case that the participants in a dispute do indeed follow the evidence, but they have in fact different evidence. This could be a result of (i) taking different background beliefs into account when trying to determine what to believe; or (ii) having background beliefs influence how they "see" the evidence. Second, scientists might disagree about which theories to accept because they started in different places: Scientist A believed very strongly that theory T is true, and the new evidence makes her even more certain. Scientist B, however, found T to be pretty implausible, and although the new evidence, shared with A, makes T a bit more plausible, it is still not enough to convince B to accept T. Let's examine these possibilities in more detail because they are crucial to the practice of science.

4.3.1 Differences in Evidential Basis

Consider two ecology students, Sam (from Brooklyn, NY) and Suzy (from Brookings, SD), studying the fauna of Yellowstone National Park. Standing at the edge of the prairie, they both see a large, bulky figure in the distance, moving slowly about. Sam believes, on the basis of that evidence, that there is a bison over there. Suzy, however, doesn't believe this. She has ostensibly the same visual evidence, but she grew up in the wide-open landscapes of the West, so she is keenly aware of the risk of making mistakes in bovine identification over large distances. Thus, given all of her evidence – the visual scene in front of her and her background belief that it is easy to misidentify distant things – it would be irrational for her to believe that the thing out there is a bison, and so she doesn't believe it. On the other hand, Sam is rational in believing that what he sees is a bison, simply because (we assume) he has no background beliefs about the risks of misidentifications. What explains the difference between Sam and Suzy is that they have different but overlapping evidential bases for their beliefs. They share the immediate visual evidence, but Suzy has a further bit of relevant data (distant bovines are hard to identify) that Sam lacks. On the assumption that one has to base one's belief on one's entire evidence, Sam and Suzy's disagreement can be traced back to differences in evidential bases. Therefore, the idea that rationality is a function of the evidence one has is not threatened.

Here's a second example, this one a bit more realistic. Two expert entomologists were standing in a field infested with a dozen species of grasshoppers. When the rancher asked if he was going to need to treat, both scientists estimated the same pest density, considered the same population dynamics, extrapolated identical developmental rates, and used the same estimate for grasshopper food consumption per day. But one entomologist said there was not a serious problem and the other concluded that treatment would be necessary. At first, this looks like a strange disagreement. However, their disagreement can be traced to differences in background data. Here's how.

The "no problem" scientist had done considerable laboratory research. He'd made wafers out of various grasses, put them in cages with a laboratory strain of grasshoppers, and weighed the wafers every 6 hours. He learned that the grasshopper fed only during the day and interpreted the evidence in the field based on these laboratory findings to draw the conclusion that forage losses would be modest.

On the other hand, the "major problem" scientist had conducted extensive field research, including having collected grasshoppers at dawn and dissected their digestive tract. He learned that some species had guts full of grass (although not the species that his "no problem" colleague studied in the lab), indicating that they fed at night. So he interpreted the evidence in the rancher's pasture based on these findings and concluded there would be a high loss of forage.

4.3.2 Theory-Ladenness of Observation

One crucial assumption behind the idea that evidence can serve as a neutral arbiter between rival theories, and thus resolve scientific disputes in a rational manner, consists in the claim that even if two people disagree in their theoretical conclusion, at least they can agree about the nature of the evidence they share. In other words, data is supposed to be theory-neutral – we can state clearly what exactly the evidence is in a way that is independent of our theoretical commitments. This assumption, however, has been questioned on various grounds, some related to scientific practice, others to the history of science, while still others draw on results of research in the cognitive sciences. Let us briefly look at some of those reasons to doubt the assumption that evidence is theory-neutral (recall also our discussion of Popper in Chapter 3).

The history of science provides some reason for doubting that the evidence can always be characterized fully independently from one's theoretical commitments. As an example, consider early research on electrical phenomena (we are talking about the late 1600s). Two observations fascinated researchers: electrostatic attraction and electrostatic repulsion – at least that is how we would describe it today. In the beginning, there was no agreement about whether repulsion is an electrical or a mechanical phenomenon. Attraction was explained by postulating fine particles of ordinary matter that, when an object like amber was rubbed, would be ejected from the object and then move through the surrounding air, sweeping up light particles, some of which ended up adhering to the amber. On this view, repulsion was simply a mechanical bouncing-off of particles from the object. Thus, these kinds of theories did not have to account for electrostatic repulsion at all. Therefore, a theory of electricity could not be criticized on the ground that it didn't explain repulsion, for repulsion was not seen as lying in the domain of theories about electricity. We will see more of these kinds of arguments when we talk about scientific revolutions and Thomas Kuhn's (1962) analysis of them, in Chapter 15.

There is plenty of research in the cognitive sciences that suggests that observation is theory-laden, which means that the evidence collected through observation is not theory-neutral. For example, research on the perception of motion shows that in addition to low-level motion detectors, background beliefs can also decisively influence what we observe. Imagine being in a movie theater. What you see is a sequence of static images projected onto your retina. If the edges of objects are displaced in the right way from static image to static image, you perceive the edge as if it was moving continuously from left to right, say. This perception is the result of how the low-level motion detector works. Now consider a pan shot of a car chase. If done right, the car stays always centered in the image, while the edges of the background objects – houses, mountains, etc. – change from static image to static image. However, it is not the case that you perceive the mountains continuously moving from right to left; rather, you see the car continuously moving from left to right. Your background belief that it is cars that move, and not mountains, influences, and in fact overrides, the output of the low-level motion detectors. This is a clear instance of the theory-ladenness of observation: What you perceive – how something looks to you – depends partly on your background beliefs.

Here is another example which illustrates how an observer having little or no theoretical background sees an event very differently than an observer with a well-developed theoretical context. Consider someone who recently moved to the US and watches his first football game. He sees how one player catches that ball-like thing out of the air after he has crossed a white line just in front of two upright, yellow poles that are connected by another, horizontal, yellow pole. His friend Jane, who attends football games every Sunday, jumps up and screams in utter joy. The US newbie is happy for her, but doesn't know what he just saw. Although both had the same visual input, only Jane, the football expert, saw it as a touchdown, while her friend did not. So for him, what he saw could not be evidence for the belief that Jane's team is winning, while the same visual input does provide Jane with such evidence. In a slogan, they saw the same thing, but didn't see it *as* the same thing. Data are evidence if they are seen *as* evidence. Experts and novices often "see" different things when they look at the same visual scene. But does this really pertain to observations of natural phenomena or scientific results?

Here's a picture (Figure 4.2) that shows something very interesting; can you guess what it is?

Your pretheoretic interpretation is almost certainly to be as unsure as the poor chap watching a football game for the first time. Perhaps this appears to be a photograph of celestial objects moving through deep space (relevant to astronomers) or maybe city and street lights photographed at night from a high elevation (relevant to human ecologists). To find out what it is, check the end of this chapter.

These examples (football and the picture you now understand after skipping ahead) illustrate the importance of theory-ladenness by contrasting an (almost) *pretheoretical* interpretation with that of a rich theoretical framework. However, these cases don't show the way in which people with *different*

Figure 4.2 What Do You See? [ME note: reproduce this in mono in print and colour in ebooks].

theoretical backgrounds can differ markedly in what they see, which gets us closer to the problem of how two scientists can disagree when presented with the same observational evidence. So consider this scenario.

In conventional rangeland grasshopper control programs, the USDA monitored aircraft applications of insecticides to make sure that none of the infested land was "skipped" – but what is the best way to understand the nature of these untreated areas in light of subsequent outbreaks, which sometimes happen? There are two major proposals.

Theory of Scorched Earth: The traditional method of application was to assure a "blanket" of treatment, with any skipped areas of high density being seen as leaving behind the "seeds" of another outbreak. Grasshoppers surviving in these unintentionally untreated swaths were viewed much like cancer cells surviving a bout of chemotherapy – they would spread into the surrounding rangeland. In fact, such lands tended to become reinfested, which was attributed to the survival of relatively high pest densities in the areas missed by the aerial applicators.

Theory of Sanctuary: The alternative view was that the untreated areas could function as refuges for natural enemies, allowing predators and parasites to reestablish control in the years following an outbreak. So, subsequent infestations of such lands were attributed to the loss of natural enemies and the rapid recovery of grasshoppers into this enemy-free space. It was the low density of natural enemy survivors in the treated areas that produced the next outbreak. Today, we intentionally leave alternate swaths of rangeland untreated to both save money and conserve natural enemies, while the immediate level of control is diminished by only a modest amount.

What this clearly illustrates is how background differences in emphasis – pest density vs. predator density – can result in different perceptions, literally. The same picture of an area can be seen by the pest-guy as "missed treatment surrounded by successful treatment" and, by the predator-guy as "predator survival surrounded by massive predator killing." This difference in how the evidence is seen can go a long way towards explaining why there are competing theories about the new outbreak.

4.3.3 Differences in Prior Probability Assignments

Sometimes scientists differ in their assessment of a given theory, not because they bring different evidence to bear on the theories, but because they differ from one another in their initial estimation of how good the theory is (e.g., that a single species of locust can change into radically different forms in response to crowding). Scientist A might initially think that some theory T really has a good chance of being true, while B might think that it's a lousy theory – T just doesn't strike her as very plausible (how many animals take on entirely different morphologies just as a result of population density, after all?) and her mentor loathed it.

This sort of difference can be made more precise by using some probability theory. In particular, we can introduce a distinction between the *prior* probability (or prior) someone assigns to a theory, and the *posterior* probability she assigns to that same theory, after she has collected more evidence. In this context, when we talk about the prior (probability) assigned to a theory or hypothesis by a scientist, we mean to express the degree of confidence they have that the theory or hypothesis is true. There is a celebrated theorem in probability theory, the so-called Bayes' Theorem, which provides a rule for how this prior should be changed in light of new evidence. If the new evidence supports the theory, the posterior probability should increase by a well-defined amount, reflecting the likelihood of the theory and the expectedness of the evidence. Now imagine two scientists evaluating the same theory, and suppose they assign different priors to that theory. In such a case, if the priors are sufficiently different, then even if the two scientists use Bayes' Theorem to change their confidence in the theory on the basis of same evidence, the resulting posterior probabilities can still be very different. Let's say that A assigns a prior of 0.3, and B one of 0.7. It would take a really huge amount of evidence to make the posteriors converge to the same number, even though both scientists take exactly the same evidence into account in exactly the same way. Thus, even with shared evidence, their disagreement about the theory remains.

Of course, this raises the question of why the scientists disagree about the priors in the first place. Above, we somewhat jokingly said that one scientist's mentor might have hated the theory, and this resulted in a low prior on part of that scientist. Another possible reason for such differences is this. Before evidence in support of the theory is collected, one might evaluate it in terms of one's background knowledge. Then, if the relevant background knowledge is sufficiently and different between scientists, their priors will be correspondingly different. We will say more about priors in the next chapter.

4.4 Conclusion

In this chapter, we have discussed different kinds of evidence and their shortcomings or advantages, from anecdotal evidence to natural histories to case studies. In Chapter 6, we will discuss experimental evidence. We have also seen that in addition to a causal theory of evidence, the so-called phenomenal theory of evidence has been proposed, which connects evidence with rationality. On this view, something is evidence for a claim if it gives a person reason for believing that claim (to a certain degree). Finally, we have identified three ways in which scientists can disagree about some theory, even though they appear to evaluate it in light of the same evidence: They might have different background beliefs, some of which are evidentially relevant to the theory at issue; they might see the evidence in different ways due to the theory-ladenness of observation; and finally, they might bring to the table different prior probabilities for the theory, which a realistic amount of new evidence cannot reconcile. And speaking of probabilities, this slippery concept is the subject of our next chapter.

P.S.: Figure 3 shows beta radiation as detected in an isopropanol cloud chamber. Courtesy https://www.nuledo.com/en/our-products, CC BY-SA 4.0, https://commons.wikimedia.org/w/index.php?curid=62073546

Note

1   The SEP entry referenced below summarizes the current discussion on peer disagreement in epistemology, which is interesting but mostly tangential to our concerns.

Annotated Bibliography

Richard Feldman and Earl Connee, 1985, "Evidentialism," in *Philosophical Studies* 48 (1):15–34.
  Classict development of the evidentialist position about how one should rationally react to the evidence in one's possession.

Bryan Frances and Jonathan Matheson, 2018, "Disagreement," *Stanford Encyclopedia of Philosophy*. Available at https://plato.stanford.edu/entries/disagreement
  A throrough presentation of many issues surrounding disagreement from a general epistemological perspective. Of particular importance for the philosophy of science is the discussion of peer disagreement.

Norwood Russell Hanson, 1958, *Patterns of Discovery*. London and New York: Cambridge University Press.
  This book contains the classic discussion of the theory-ladenness of observation, including a discussion of the distinction between seeing-as and seeing-that.

5

THE NATURE OF PROBABILITY

5.1 Basics of Probability

Here is what we need to know about probability theory for our purposes. First, a probability (abbreviated as "pr") is simply a number that attaches to *propositions* (the contents of declarative sentences) in rough proportionality to their chance of being true, or to *events* in rough proportionality to the chance of their occurrence. Those numbers lie between 0 and 1, such that 0 is reserved for propositions that cannot possibly be true and 1 for those that cannot possibly be false. For example, "the nonzero integer n is both odd and even" cannot possibly be true. Thus pr(n is both odd and even) = 0. On the other hand, pr(n is odd or even) = 1.

Now consider any nonzero integer n chosen at random. What is the probability that it is even? Clearly, pr(n is even) = 0.5, because being even is one of only two options, and we have no reason for believing that the randomly chosen number is somehow special. Thus, in this case at least, the probability of being even equals the frequency of even numbers among all nonzero integers.

Let's move on to examples involving games of chance, which were the original focus of probability theory. What's the probability that a card drawn at random from a standard deck of cards is an ace? There are 52 cards in a standard deck, four of which are aces; thus, the probability is 4/52 = 1/13 ≈ 0.08. Suppose you want to know next what the probability of a red ace is? There are two red aces in a standard deck, which makes the probability equal 2/52 = 1/26 ≈ 0.04. Now imagine that you know that a card that has been drawn is red. What's the probability that its suit is "heart"? There are 26 red

This is Philosophy of Science: An Introduction, First Edition. Franz-Peter Griesmaier and Jeffrey A. Lockwood.
© 2022 John Wiley & Sons, Inc. Published 2022 by John Wiley & Sons, Inc.

cards in the deck, 13 of which are hearts, which means that the probability of having drawn a heart, given that what has been drawn is red, equals 13/26 = 1/2. We write this as follows, with "H" standing for "being a heart" and "R" for "being red": pr(H/R) = 1/2 = 0.5. Notice that this *conditional probability* is bigger than the simple probability of drawing an H. The latter is pr(H) = 1/4 = 0.25. We can define conditional probabilities in the following way (assuming of course that the denominator, pr(R), is > 0):

$$\text{pr}(H/R) = \text{pr}(H \& R)/\text{pr}(R) \quad \rightarrow\rightarrow \qquad (\text{CPr})$$

The probability of (H & R) – of being a red heart – is of the course the same as the probability of being a heart (since all hearts are red) and thus equals 1/4. Since pr(R) = 1/2, pr(H&R)/pr(R) = 1/4/1/2 = 1/2. This agrees with the value we have determined earlier for pr(H/R), which illustrates (but of course doesn't prove) the correctness of our definition of conditional probabilities.

Now we are in a position to derive a theorem important in many discussions about the role of probabilities in science, the so-called Bayesian theorem (or Bayes' Theorem; Thomas Bayes was an eighteenth-century English mathematician, philosopher, and minister). By simple arithmetic, using more general variables A and B, we can transform the definition of conditional probabilities into the following equation:

$$\text{pr}(A \& B) = \text{pr}(A/B)\text{pr}(B) \qquad (5.1)$$

Similarly,

$$\text{pr}(B \& A) = \text{pr}(B/A)\text{pr}(A) \qquad (5.2)$$

Obviously,

$$\text{pr}(A \& B) = \text{pr}(B \& A) \qquad (5.3)$$

[being a heart and red is the same as being red and a heart].
Therefore, from (1), (2), and (3), it follows that

$$\text{pr}(A/B)\text{pr}(B) = \text{pr}(B/A)\text{pr}(A) \qquad (5.4)$$

Dividing both sides by pr(B), we get
Bayes' Theorem:

$$\text{pr}(A/B) = \text{pr}(B/A)\text{pr}(A)/\text{pr}(B) \quad \rightarrow\rightarrow \qquad\qquad (BT)$$

In the context of discussing evidence, the theorem is usually written in a slightly different way, with "H" standing for "hypothesis" and "E" for "evidence":

$$\text{pr}(H/E) = \text{pr}(E/H)\text{pr}(H)/\text{pr}(E)$$

This simply says that the *posterior probability* (so-called because it comes *after* the evidence is taken into account) of some hypothesis H, given evidence E [the term pr(H/E)], equals the product of the *likelihood* of the hypothesis [pr(E/H)] and its *prior probability* [pr(H)], divided by the *expectedness* of the evidence [pr(E)]. Be careful when reading this: The forward slash between propositions, as in pr(E/H), indicates a conditional probability (the probability of E, given H), while the same forward slash between probabilities, as in pr(H)/pr(E), indicates division or a ratio of probabilities. Consider our earlier case of the welder's apparently fume-related cancer:

pr(fumes cause cancer, given the welder has cancer) =

pr(the welder has cancer, given the fumes cause cancer) x pr(the fumes cause cancer)/pr(the welder has cancer)

5.2   Interpretations of Probability

It is widely agreed that our term "probability" might actually cover three importantly different concepts. (i) First, there are purely subjective degrees of conviction, or *credences*, that might be represented as probabilities. For example, you might strongly believe that the Denver Broncos will win next year's Super Bowl and be willing to bet $80 for a chance to win $100. Beginning with Bruno de Finetti (an Italian statistician who worked in the first half of the 1900s), various scientists/philosophers have argued that this particular willingness of yours to bet $80 for a chance to win $100 suggests that your credence that the Broncos will win is 0.8. This latter notion of a credence has been used in analyzing how a scientist should change her confidence in the truth of a theory as new evidence comes in. (ii) Next, there is an *epistemic* notion of *probability*. It simply assigns a value to the chance that a proposition is true or that an event will occur, given a certain body of evidence. For example, there is a certain probability that a person has a particular disease, given that they tested positive for the disease. That probability is not 1; tests have false positives, and it is also necessary to take the base rate of this disease into account (i.e., the probability that any person has the disease). (iii) Finally, there is probability as an *objective chance*. The chances that a smoker develops lung cancer are an objective matter of fact and independent from what people believe about those chances.

5.3   Probabilities as Credences

The starting point for this approach is the observation that beliefs do not seem to be an all-or-nothing affair – I either believe or I don't. Rather, beliefs come in degrees – I believe some things more strongly than others. For example, I might believe more strongly that it's going to rain later today than I believe that the store ran out of my favorite cookies, although I may believe both things. Similarly, scientists believe some theories more strongly than others, even before they have evaluated any relevant evidence, in light of their background knowledge. For example, it will make some theories appear more plausible than others, which will be reflected in a belief about a theory's probability. Such background knowledge therefore influences the scientist's assessment of the theory, and the results of those assessments are credences – how strongly the scientist believes that theory. Purists in the credence camp insist that it actually doesn't matter exactly how background knowledge leads to initial credences, or prior probabilities. However, in order for such initial credences to be rational, they must respect two constraints: First, the credences must be *probabilistically consistent*, and second, credences have to be updated in light of new evidence by a process called *conditionalization*. Let's look briefly at those two constraints.

5.3.1   *Probabilistic Consistency*

Credences are *probabilistically consistent* if they obey the probability calculus. For example, suppose one had a credence of 0.6 that it will rain within the hour, and a credence of 0.6 that it will not rain. Clearly, this is probabilistically inconsistent, because the credences attached to a proposition and its negation (rain and no rain) need to add up to 1. You might think that it is always easy to respect the consistency constraint, but in practice, it might be difficult to see that two propositions are actually the negations of each other so that the credences assigned to them should add up to 1. For example, "It is not the case that if it rains, the streets remain dry" is the negation of "It either doesn't rain or the streets remain dry." We are sure you have to think about it for a bit to see that one is the negation of the other.

Why should we be probabilistically consistent? The short answer is this: If a person's credences are not probabilistically consistent, she runs the risk of accepting a set of bets on some event, say a horse race, such that no matter who actually wins the race, the person who bets will lose money (such a set of bets is called a *Dutch Book*). Since she will lose no matter what, she'd be irrational in accepting such a set of bets. But if she has inconsistent credences, she is disposed to accept them, and thus disposed to act irrationally.

5.3.2   Conditionalization

Conditionalization is the process of changing one's credences as new evidence comes in by conditionalizing those credences on the evidence. If at some time t one assigns pr(H) = v (for some v between 0 and 1), then one ought to assign a probability to H at some later time t*, when new evidence has been obtained, in such a way that pr(H) at t* = pr(H/E). How we calculate pr(H/E) is the content of Bayes' Theorem, which we discussed earlier. Some authors have argued in support of the conditionalization constraint by trying to develop a *Diachronic Dutch Book* argument, which is supposed to show that if one doesn't update credences by conditionalization, one is subject to a temporal sequence of bets with a guaranteed loss. The idea is the same as above: Accepting a sequence of bets that guarantee a loss, no matter what, is irrational, and thus it is equally irrational to update in a way that disposes one to accept such a sequence.

5.3.3   The Problem of Priors

There is a deep issue regarding priors. And this problem is also a potential source of scientific disagreement. Recall that *priors* are the prior probabilities (or prior credences) someone assigns to a hypothesis before new evidence comes in. If the only constraint on priors consists of probabilistic consistency (conditionalization is a constraint on posterior probabilities), then two or more people can have radically different priors. Take the example of one's prior credence about rain (R) or not rain (¬R) later today. John might assign pr(R) = 0.6 and pr(¬R) = 0.4. Since both add up to one, this set of credences is consistent. Jane also has a consistent set, namely, $pr(R) = 0.1$ and $pr(\neg R) = 0.9$. But obviously, John and Jane have radically different credences. Who is right? According to the purist, there is no answer to this questions. Purists are subjectivists about probabilities. You can have any set of credences you like as long as they are consistent and you update by conditionalizing.

So do we achieve agreement by conditionalizing on the same evidence? That would seem to be ideal for science because we'd all eventually converge on the same posterior probability. However, the answer is a mixed bag. There are various theorems showing that in the long run, no matter how far apart the initial priors are, the posteriors, after repeated instances of updating on the same evidence, will eventually converge on the same value. However, the long run is often a very long run, so long in fact as to be impractical in many situations in which good probability estimates are important, as might happen with drug trials in medical research. At any rate, the problem of priors is one important source of scientific disagreement, because if the priors attached to a theory by two or more scientists are sufficiently far apart, then even when they conditionalize on the same evidence, they will disagree about the posteriors as well, at least in the time-frame characteristic of much of scientific research. Thus, casting disputes about scientific theories in terms of credences allows us, together with the formalism of Bayesian updating, to get a clear sense of why disagreement might persist even after a lot of relevant evidence has been collected and evaluated by all parties in the dispute: Large differences in priors translate into sizable differences in posteriors, which means chronic disagreement.

Clearly, the problem results from the fact that there are no constraints on prior credences other than consistency. Maybe we could make some progress if we were to introduce further constraints that can reign in the promiscuity about priors. What could such constraints be? Philosopher Abner Shimony has proposed a couple of rough-and-ready additional constraints. For example, if you have no evidence at all that either supports or undermines a particular claim (hypothesis or theory), your initial credences should be around 0.5. Another proposal is to avoid assigning extreme probabilities, such as 0.001 or 0.999, before the evidence is in. One widely discussed additional constraint derives from the so-called *principal principle*, originally proposed by David Lewis in 1980, according to which, roughly, one's credences should reflect known chances. My credence that a fair cutting of a deck of cards producing a heart should be 0.25, because I know that the chance of a card being a heart is 0.25. This means that credences are transformed into *epistemic probabilities* – probabilities we assign in light of evidence. We discuss those below. But there is another troubling problem with the credence approach to probabilities that is particularly relevant to the philosophy of science. It is the problem of old evidence.

5.3.4   The Problem of Old Evidence

In order to clearly see the problem, we need to return to Bayes' Theorem and make the role of background knowledge explicit, because whenever we evaluate a theory, we also take into account our background knowledge. Formally, this is easy to do – we simply conditionalize every term in the theorem on a variable, K, that stands for background knowledge. The theorem now looks like this:

$$\mathrm{pr}\big(H/E\,\&\,K\big)=\mathrm{pr}\big(E/H\,\&\,K\big)\mathrm{pr}\big(H/K\big)/\mathrm{pr}\big(E/K\big) \qquad \big(\mathrm{BTK}\big)$$

Now let's suppose that a new theory, H, has been published, which accounts for certain phenomena rather elegantly. Then it turns out, to everybody's surprise, that it also accounts for a phenomenon that had long puzzled the scientific community. Realizing this surprising fact, the theory gets a strong boost – people believe it even more strongly now than they did before they realized that it solved that pesky old problem. After all, H was not designed to even address that problem. Can this increase in credence be captured by (BTK), above? Well, it seems the answer is a resounding no. The old problem, call it E, has long been part of the background knowledge K of the scientists involved. This means of course that pr(E/K) = 1, because if E is part of K, then the probability of E, given K, must be 1. Similarly, pr(E/H & K) = 1, because again, E is already part of K on which the likelihood conditionalizes. Thus, the expectedness and the likelihood both equal 1 and therefore drop out of the equation. What's left is pr(H/K), which is the prior attached to H. Thus, the posterior probability pr(H/E & K) = pr(H/K), which is exactly the prior of H (in this case, prior to the realization that H neatly solves the old problem). Realizing that H accounts for that old problem does not result in any credence boost!

The fact that we introduced the problem by talking about a realization that the new theory solves an old problem is just one form of a more general problem. *Any* old evidence, whether used in formulating the theory or not, is such that its credence-boosting power cannot be captured using the Bayesian theorem. Just stop for a moment and let this sink in: If a scientist uses known data in formulating her theory, the fact that the theory accounts for those data should not increase our credence in the theory. And of course most, if not all, theories are in fact formulated using known data. Thus, the data can't do anything to provide any credence to the theory beyond the initial plausibility that it has. Something is wrong here.

The radical solution to this problem is to give up on the credence approach altogether. A less radical solution that has received a good deal of attention in the literature is to "go counterfactual." We will discuss at length what counterfactuals are in the next chapter. For our current context, we can usefully identify them with *pretending*. In assessing the evidential value of known pieces of evidence, we *pretend* that we don't know them, assess the probability of the theory, and then see what happens if we add these pieces of evidence back in by applying Bayes' theorem. In formal terms, we "subtract" E from K, determine the probability of H conditional on K-E, and then conditionalize on E to see how the probabilities change. There are interesting technical questions connected with this kind of solution to the problem of old evidence, which we cannot discuss here. However, there are strategies in model validation that bear a striking similarity to this sort of "pretending" solution.

Suppose we have a limited set of data about some ecosystem, going back maybe 40 years, and we want to model the relations between tree density and temperature in a forest as a function of time. If we were to use all of those data in constructing our model, we might have to wait for a fairly long time before we can actually test the model, perhaps as long as several years, if not decades. To avoid this problem, we first *pretend* not to have some of the data, thus ignoring them for the purposes of model construction, and *then* we test the resulting model by adding back in the data that we have ignored during the construction. In this way, we mimic the prediction of new data, something we usually use in model validation.

The issue this technique addresses is often identified as that of accounting for the difference between *data accommodation* and *data prediction*. The thought is that a model that simply accommodates the data gets less confirmation from those data than a model which can also predict new data. Accommodation is seen to be more easily accomplished than successful prediction. For example, laying a curve through already known data points is fairly easy. But finding the curve that also correctly predicts new data is much more difficult.

5.4   Epistemic Probabilities

Roughly speaking, epistemic probabilities measure the degree of confidence one should have in the truth of a proposition relative to a body of evidence. In our discussion of the credence approach, we already encountered such probabilities – they are Bayesian posterior probabilities, which are probabilities of propositions conditional on evidence. They don't measure any objective chance of an event happening, such as the chance of a cesium atom decaying during the next 4 hours. But they are also not expressions of mere subjective convictions, such as prior credences are on the purist approach discussed earlier. Rather, epistemic probabilities tell us how much a claim (theory or hypothesis) is supported by the evidence one has. We already discussed how such probabilities can be calculated using Bayes' theorem.

5.4.1   The Classical Interpretation

A special case of epistemic probabilities arises in the context of games of chance when the evidence doesn't favor one possible outcome over any of the others. It was in this context that the theory of probability was originally developed. The accompanying interpretation is therefore called the *classical interpretation*. Consider throwing a fair die. What is the probability of throwing a five? To answer this question, we need to determine the so-called *probability space*, which is in this case simply the set of all possible outcomes of throwing a die. Clearly, with a standard die, there are six possible outcomes. These constitute the *probability space*. Now, throwing a five is the actualization of one of those six possibilities. If we have no information that would suggest that the die is somehow rigged or biased, we have no reason for believing that one of the possibilities has a better chance of being actualized than any of the others. In other words, we have no evidence for one possibility over the others. In this situation, the classicist simply says to treat all outcomes as equally probable. This advice is codified as the *indifference principle*: If there is no reason for thinking that one outcome is privileged over the others, one should be indifferent as to which outcome will materialize. Such indifference means that throwing a five has a probability of 1/6, because there are six possibilities among whose probabilities we should be indifferent, and because the probabilities have to add up to 1, $pr(1) = pr(2) = pr(3) = pr(4) = pr(5) = pr(6) = 1/6$.

This is a special case of epistemic probabilities in the following way. $pr(6/E) = pr(5/E) = $ etc., because the evidence is the same for each possible outcome, in the sense that whatever evidence we happen to have in our background, none of it gives us a reason for thinking that one possibility is privileged over any of the others. In other words, conditionalizing on nonprivileging evidence results in equal probability assignments to every element of the probability space. We can also think of the evidence relevant to the outcomes of throwing a die as the information that the die is unbiased, or fair. Conditionalizing on the evidence that the die is fair, each possible outcome receives the same probability; $pr(4/Fair$ $Die) = pr(2/Fair Die) = $ etc. The evidence doesn't distinguish among the possible outcomes, so all of them are equiprobable relative to the evidence that the die is fair.

In ecology, organisms tend to be spatially dispersed in either clumped or uniform patterns. But there is a case in which the classical interpretation of epistemic probabilities yields an expectation just like rolling a die. In a random distribution, as seen when organisms are in a homogenous environment, lack any social interactions (either positive or negative), and are spread by some abiotic factor such as the wind or ocean currents, each location is equally likely to be occupied. So, the spatial pattern of wind-dispersed dandelion seeds is equivalent to the pattern seen in a temporal pattern of coin flips.

5.4.2   Bertrand's Paradox

Many of us are familiar with the classical interpretation of the epistemic notion of probability. If you studied any probability in high school, examples from games of chance together with this classical interpretation were most likely used. And yet, there is an important paradox connected with the classical interpretation, known as *Bertrand's Paradox*, named after Joseph Bertrand who discussed several examples of it in the late nineteenth century. At bottom, the paradox shows that there are cases where the application of the indifference principle leads to contradictory probability assignments. This can happen when the same outcomes are describable in mathematically equivalent ways. Following Bas van Fraassen's exposition, here is a simple case illustrating the paradox.

Imagine a factory that produces, with no errors, metal cubes with edges of a length anywhere between 0 and 1 inches. (Obviously, a "cube" with 0 edge length is not a cube. But as you'll see, this simplification of the problem doesn't really matter.) Assume further that it doesn't produce more of one size of cubes than of any of the others. What's the probability that the next cube it produces has an edge length of less than ½ an inch? Applying the principle of indifference, the answer should be 0.5, as there is no reason to privilege the lower half of possible lengths over the upper half. Now let's ask the same question rephrased in an equivalent way: What's the probability that the next cube it produces has a face area of less than ¼ of a square inch? Applying the principle of indifference again, we get the result that the probability is 0.25. The probability space under this description of the case contains four possibilities: a face area between 0 and ¼, between ¼ and ½, between ½ and ¾, and between ¾ and 1. Four possibilities, and since we have no reason for privileging one over the other, the relevant probability should be 0.25. But this is disastrous! A cube with a face area between 0 and ¼ of an inch is the same kind of cube as one with a side length between 0 and ½ of an inch! The same possible outcome is assigned two different probabilities by the principle of indifference. Something's gone seriously wrong.

Some have attempted various technical fixes, the details of which lie beyond the scope of our brief survey. Others have argued that perhaps the Bertrand paradox reflects a deeper problem with the classical interpretation, namely, that we are trying to use ignorance, codified in the indifference principle, to determine probabilities. Maybe in cases in which we truly don't have access to any relevant evidence, we should abstain from assigning any probabilities at all. Perhaps it is this kind of sentiment that's most explicitly embodied in the final interpretation of probability we discuss here, the so-called *frequency interpretation* of the notion of objective chances.

5.5   Probabilities as Objective Chances

We mentioned earlier that there is a concept of probability that is supposed to capture the objective chance of some event happening, quite independently of how convinced anybody is about its chances of happening and also independent of the actual evidence (including background knowledge) that anybody has relevant to a hypothesis about what those chances are. In other words, those objective chances are different from both credences and from epistemic probabilities. They are just out there. But if so, is there any way to find out what those chances are?

5.5.1   *Frequentism*

The so-called *frequentists* believe that the answer is a clear yes, and that we can determine those probabilities purely empirically, without having to rely on the indifference principles. We can determine those probabilities by observing *relative frequencies* (thus the name "frequentists"). Suppose you want to know for a particular coin what the probability of *heads* is. What you do is simple: You toss the coin over and over again, diligently recording the number of heads. After a long sequence of tosses, you calculate how often the coin came up heads. This number is the relative frequency of *heads* in an actual sequence of tosses. And this number *is* the probability of *heads* for this coin.

It is clear on just a moment's reflection that this definition of probabilities as the actual relative frequency of the target property is problematic. Suppose, as I walk down a street, I toss a coin once, it comes up heads, but after I toss it again, it falls into the gutter and is lost forever. Should I now conclude that the probability of *heads* with this coin equals 1? Clearly not. This problem has prompted statisticians to define probabilities, not in terms of actual frequencies, but instead in terms of the *limit of the observed frequencies* in a potentially infinitely long trial. The hope is that in such a trial, idiosyncrasies that might beset short trials will wash out. Unfortunately, this doesn't put all worries to rest. First, there is simply no mathematical guarantee that an infinitely long trial will produce a sequence of ratios that converges to a unique limit. If it doesn't, then there would be no probability at all assignable to the target property. Second, even if there is such a unique limit, no finite segment of the infinite sequence guarantees that the observed frequency actually coincides with the limit of such frequencies in an infinitely long trial. Observing a relative frequency of 0.5 heads in trial consisting of 1,00,000 tosses is compatible with the limit being 0.33. Anything could happen after the first 1,00,000 tosses. Just consider playing roulette and having red come up 20 times in a row. Surely, this could have been it for red, and the next 200 times the croupier rolls the ball, it lands on black. Equally surely, it could be the case that the red streak continues forever. We just have no way to know, precisely because the events are statistically independent from each other. Thus, even the limits of frequencies do not provide us with a good definition of probability. On the one hand, no such unique limit may exist (a metaphysical problem), and on the other hand, even if a limit does exist, the finite data we have will never assure us of what that limit is (an epistemic problem).

Maybe the problem is that we want too much – *assurance* of knowing the unique and correct numbers. Maybe we could be somewhat less demanding by relating the use of probability to our notion of evidence as reason giving. From such a perspective, actual frequencies provide us with a *defeasible* reason for believing that the probabilities are such-and-such. That we run into the epistemic problem mentioned above simply shows that actually observed frequencies are not conclusive reasons for beliefs about the relevant probabilities. Before we explore the details of this perspective, we need to mention another problem for the frequentist and the proposed solution in terms of so-called *propensities*.

5.5.2 Propensities

Frequentism faces the problem of single-case probabilities. For example, the half-life of carbon-14, a radioactive isotope used in carbon dating, is 5,700 years. That means that each particular C-14 isotope in a sample has a 50% chance of disintegrating within the next 5,700 years. Clearly, this probability of 0.5 is not the limit of any frequency. If a particular isotope disintegrates, then that's that. So, what is this sort of probability?

Popper, among others, proposed to identify this probability with a *propensity*, a sort of strong causal tendency, or disposition, to manifest a certain behavior. It is somehow part of the nature of a C-14 isotope to have a 50% chance of manifesting disintegration within 5,700 years. It also has a 50% chance of manifesting nondisintegration. To many, this approach sounded quite mysterious. In particular, it seems to be in conflict with the principle "same cause, same effect." The nature of the C-14 isotope, which grounds its behavior, sometimes gives rise to disintegration, and sometimes it does not. In addition, the propensity interpretation faces a number of technical problems. But we will leave it at that and return to the idea that the notion of defeasibility might help in dealing with probabilities.

5.6   Probabilities and Defeasible Reasoning

Instead of despairing about actual relative frequencies, because they don't deliver determinate probabilities, we suggested looking at those frequencies as providing us with defeasible reasons for certain probability assignments. This means that if you observe 60% heads in a long trial with a coin, you have a defeasible reason that the probability of heads for this coin is 0.6; of course, if you continue your trial, you might acquire defeating information, such as pronounced shifts in the relative frequencies of heads and tails. Such a switch in perspective is also advisable in light of the ultimate motivation behind a frequentist approach, which derives from an allegiance to empiricism. An empiricist holds that all knowledge must ultimately be grounded in observation, which means that alleged knowledge of probabilities must also be so grounded. What we can observe is relative frequencies in actual trials. Of course, any consistent empiricist will also hold that observed evidence can never provide us with conclusive reasons for our beliefs; at best, we get defeasible reasons. But once we remind ourselves of this fact, both the metaphysical and the epistemic problem about probabilities as frequency limits vanish. Our observations constitute defeasible reasons for our probability assignments – perhaps defeasible reasons for believing that there is such a limit and defeasible reasons for believing that this limit has a particular value.

Background knowledge can also result in defeaters. Thus, if we observe an initial run of 4,000 heads in a row, then, given beliefs about fair coins and fair tossing, we should become suspicious. Those background beliefs can become defeaters for the assumption that the coin is indeed fair. Given the evidence (observed relative frequency), we have defeasible information that the probability of heads for this coin is close to 1, but (i) we might have to give up our belief that it is a fair coin, or (ii) we might contend that extending the trial will expose the long run of heads as a fluke.

It remains to be seen whether emphasizing the role of defeasibility can also move us some way toward addressing the problems we identified in connection with credences (unconstrained priors and old evidence) and the classical interpretation of probability (inconsistent probability assigments as the result of equivalent, but different specifications of the probability space). It is fairly easy to see that those problems are similar to the one besetting the frequency interpretation. Setting aside the problem of old evidence for the moment, we are worried in the remaining two cases about the absence of a fully *determinate* probability, just as we are in the case of frequency limits. Subjective credences are not fully determinate, as they are subject to the "whims" of the probability-ascribing agent. Classical probabilities are not fully determinate, as they are subject to the choice of one specification of the probability space over equivalent ones (in our example, one of the specifications partitioned the space in terms of side-length, while the other did so in terms of face-area).

How does defeasibility help? For subjective credences, recall Shimony's advice to assign a probability to a theory near 0.5 in the absence of any discriminating evidence, and also to avoid assigning extremes. These of course are defeasible rules and could be made obsolete by relevant information that acts as a defeater. You might hypothesize that the probability of heads for this coin is 0.5, but this hypothesis could be defeated by actual frequency information. Or you might even be in possession of information that steers you away from assigning 0.5 before the start of the actual trial; maybe you are told the coin has been rigged. This is perhaps a simple example of Lewis' *principal principle*: Your assignments of credences should reflect actual frequency information to the extent to which it is available.

For classical probabilities, perhaps the principle of indifference can be seen as another defeasible rule for assigning probabilities. If there is no discriminating evidence, and if the probability space does not allow for equivalent descriptions that lead to inconsistent probability assignments, it is reasonable to identify the probability of each outcome with the ratio of the outcome over the number of all possible outcomes. Should you receive information that the probability space can be redescribed in the relevant way, perhaps the reasonable course of action is to abstain from assigning a probability altogether, noticing that such cases are going to be quite rare. Sometimes, the evidential situation is such that it is simply not reasonable to make even probability judgments, just as there are evidential situations where one should withhold full belief.

This leaves us with the problem of old evidence. Switching to an emphasis on defeasible reasoning doesn't appear to be of help with that problem. After all, the very process of Bayesian updating (plugging relevant numbers into Bayes' Theorem) already embodies defeasible reasoning, as the posterior probabilities are defeasibly justified by one's current body of evidence, but can easily change as new evidence comes in. This problem might really have to be addressed in terms of counterfactual reasoning by pretending that certain evidence was not available in the construction of the theory.

5.7 Fallacies

We conclude this chapter with a brief reminder of two pervasive fallacies in probabilistic reasoning. Suppose you play roulette in a casino, betting on colors only. After observing a long streak of red, you reason that now would be a good time to bet all of your chips on black. After all, since both black and red have a near 50/50 chance of being "selected" by the ball, the red streak simply can't go on. If you reason like this, you commit the *gambler's fallacy*. You mistakenly fail to realize that the consecutive throws of the little ball are statistically independent from each other. This means that for each throw, the chance for black to appear are near 50/50, as is the chance for red. The observation that over the last 50 throws, only red appeared, does not change this fact. Compare the roulette situation with repeated draws of colored balls from an urn without replacement, and suppose you know there is an equal number of red and black balls in the urn when the drawing starts. In this case, if you have observed a large number of red balls being drawn, you should change your probability assignment of the next ball being black, because there are now many more black balls than red ones in the urn. But this sort of statistical dependence is absent when playing roulette.

The second fallacy is the so-called *base-rate fallacy*. Imagine the following scenario, borrowed from psychologists Daniel Kahneman and Amos Tversky. There is a car accident on Main Street, involving a taxi cab. The city has two cab companies, one with green cars, which make up 85% of the total, and one with blue ones, accounting for the remaining 15%. The police received a witness report that the cab was blue, and it has been determined that the witness correctly identifies, in dim lighting conditions that were present during the accident, the two colors 80% of the time, and makes an incorrect identification in 20% of cases. What is the probability that the cab involved in the accident was indeed blue, as the witness claimed?

Most people judge this probability to be about 0.8; after all, the witness is 80% reliable. However, this estimate commits the base-rate fallacy – it does not take into account the fact that there are many more green cabs than there are blue ones. The actual probability is approximately 0.41, which can be shown by using Bayes' Theorem. If you estimated the probability to be around 0.8 as well, don't be alarmed – most people commit this base-rate fallacy. However, now that you are aware of it, you can avoid it, including when you do your scientific work. For us, the base rate fallacy will become interesting again in Chapter 12, where it plays an important role in assessing a popular argument for scientific realism.

5.8   Conclusion

As is so often the case in the philosophy of science, what might have seemed to be a simple concept with an obvious meaning turned out to be far more challenging when we started asking questions. But the importance of probabilistic reasoning in science demands clarity as to exactly what we mean when saying that an event or a statement is "probable." Perhaps we are making a claim about objective chance, or maybe we are asserting something about what we know based on evidence, or it could be that we are expressing a degree of subjective conviction. Having explored the nature of probabilistic reasoning from evidence earlier, we next turn to the ways in which scientists acquire evidence, beginning with the concepts of experimental treatments and controls.

Annotated Bibliography

Alan Háyek, 2019, "Interpretations of Probability," *Stanford Encyclopedia of Philosophy*, available at https://plato.stanford.edu/entries/probability-interpret

Accessible discussion of different interpretations of probability, including much more detail on the propensity interpretation and an interesting section on the "best-system" interpretation, which we didn't cover in our chapter.

Collin Howson and Peter Urbach, 2006, *Scientific Reasoning: The Bayesian Approach*. Open Court: Carus Publishing Company.

In parts fairly technical, this book applies Bayesian probability theory to problems in the social and physical sciences. It also discusses classical estimation theory and the approach to significance tests championed by Neyman and Pearson. The authors mount a spirited defense against many criticisms of the Bayesian program.

D. H. Mellor, 2005, *Probability: A Philosophical Introduction*. London and New York: Routledge.

Assuming no mathematical background, this book provides a very accessible tour through the world of philosophical questions raised by the concept of probability, keeping technical details to the absolute minimum.

6

DO NOT BE MISLED: CONFOUNDS AND CONTROLS

6.1 Trials and Errors

Arguably the most simplistic form of an experiment is the "what happens if...?" sort of venture. In a sense, children experiment with the world in this manner on a frequent basis. Imagine a kid receives a chemistry set for Christmas. She's delighted, of course, and follows the instruction manual with grave diligence. But having observed chemical reactions according to the directions, she begins to wonder, "What happens if I combine this and that?" So, she starts mixing reagents and most of the time nothing much occurs, but then there is a high exothermic reaction which breaks the test tube and spills chemicals all over the card table. She hurriedly cleans up the mess and concludes that putting those two chemical together yields a great deal of heat. Has she conducted an experiment?

We might contend that the child's unsystematic mixing of materials wasn't anything like a designed experiment conducted in a chemistry laboratory. But conceptually, is there an essential difference? She can't explain what happened, but she infers quite reasonably that the two chemicals were inert when left on their own, that putting them together generated a reaction, and that doing so again would likely yield a similar result. The separate chemicals prior to the reaction represented a kind of control and the mixing constituted a treatment. Of course, this sort of experimentation is not the ideal in science, although perhaps tinkering is rather more common than we might suspect. In any case, we have the basis for a form of inquiry that lays the groundwork

This is Philosophy of Science: An Introduction, First Edition. Franz-Peter Griesmaier and Jeffrey A. Lockwood.
© 2022 John Wiley & Sons, Inc. Published 2022 by John Wiley & Sons, Inc.

for experimentation. What is needed for systematic and rigorous experiments is a more sophisticated understanding of treatments and controls.

6.2 Treatment and Control

Why are controls important? Consider a pharmaceutical company that wants to market a new pill for pain relief. There are several things that have to be shown before the FDA approves new medications. First, the pill needs to be shown to be effective, and second, it also needs to be shown that it is safe, meaning that it doesn't have unacceptable side-effects (e.g., causing brain tumors – a bad tradeoff for relieving a headache). But what does it mean for a pill to be effective? Clearly, taking the pill must improve the condition for which it is used faster or more often than for those people who don't take the pill. However, as has become clear over the centuries, humans are subject to the so-called placebo effect, in which an inert substance, when taken under the assumption that it is actual medication, has a positive effect on a patient's recovery. Thus, two things need to be determined. First, does the treatment result in better recovery (faster, or more frequently) than the nontreatment (negative control)? Second, does the treatment result in better recovery than the placebo (positive control)? We'll not address the matter of the side-effects, but the same line of reasoning pertains – does the new pill result in the occurrence of brain tumors at a rate significantly different than the negative and positive controls?

These two questions regarding comparisons to controls are not confined to the contexts of clinical trials, where we want to determine the safety and effectiveness of a drug. Negative and, to some extent, positive controls are also involved in many other experiments. Here are some examples.

(1) Suppose an entomologist wanted to determine if a new chemical is toxic to insects. And let's assume the chemical is soluble in alcohol. The researcher could apply the compound dissolved in alcohol to the body of some insects and leave others untreated (negative control). But if half of the former and none of the latter died, then can the entomologist be certain that the alcohol wasn't a factor? After all, alcohol could dissolve the waxy cuticle of the insect and thereby cause some level of mortality by desiccation (water loss). So a valid conclusion would require applying just alcohol (positive control) to the insects – and if half of the insects died then it would be reasonable to conclude that the

new chemical was not a promising insecticide. If, on the other hand, the carrier had seemingly no effect (none of the insects died as a result of the application of alcohol alone), then it would be reasonable to use only negative controls in the next step of the study. In this step, the scientist would want to study the effect of the insecticide when used on pests in some crop, and leaving some fields untreated constitutes such a negative control. A positive control, i.e., spraying some fields with alcohol only, would seem pointless.

(2) Suppose a psychologist wanted to determine if "priming" people with an image of a human infant would make them more likely to act kindly. Let's imagine that the experimenter showed subjects photographs of babies and then later contrived a situation in which the individuals were approached by a panhandler asking for change – and half of the subjects gave money. The problem here is whether any picture of another person or any photo of a baby creature would prime generosity. So the experimenter might use positive controls including photos of adult humans and puppies. And if those who saw puppies gave at rates equal to those who saw babies, the conclusion might be that images of young or vulnerable beings of any species primed people for kindness.

There is of course a connection between the desire for controls in order to rule out confounding variables, and the interference-based nature of experimentation. In many cases, we have to produce the controls ourselves. In drug trials, we gather together a group of people who are all suffering from headaches. Then we hand out the new pill to some patients, a sugar pill (with identical shape and color) to others, and nothing at all to still others. The controls are put in place by us in virtue of dividing a population into groups (treatment, positive, and negative controls) and then distributing the pills accordingly (real, sugar, or nothing) – and monitoring the reported level of pain within the three groups. This constitutes a clear case of interference.

One might accordingly think that experiments necessarily involve interference, because without it, there would be no controls. But there are important exceptions. Let's suppose that the scientist who was seeking a new pain medication began with a trip to the Amazon to learn how the indigenous people relieve pain. Our researcher knows that aspirin's active ingredient is acetylsalicylic acid and that ancient Egyptians extracted salicin (a related chemical) from the bark of willows to relieve pain – so she figures that maybe an ethnobotanical approach in the Americas would be fruitful. When she arrives in a remote village with her research assistants, they accidentally step into mounds of aggressive ants that deliver dozens of excruciating bites. The village elder takes them into his house and gives them a drink of a bitter tea that he explains comes from the leaves of the "relief plant." Within minutes their pain is gone! But does this constitute an experiment? What would be the control?

We can think of the condition of the scientists prior to drinking the tea as a temporal control for the treatment. In general, temporal controls are often used in *counterfactual reasoning* about the cause of some event (in this case, the relief of pain). We claim that some event E (a change in a system) was *caused by* some factor F just in case we think that it is true that *had F not occurred, E would not have occurred either*. These counterfactual claims are a mouthful, and we will unravel them momentarily. For now, just notice how they are involved in using temporal controls, and thus in experimentation. We compare an earlier state of a system with its current state and credit the change to some event that separates the earlier state from current state. Consider the Amazonian expedition. In a sense, the earlier state of pain is a control for the treatment (the tea), because we can say that had there been no tea, the pain would not have suddenly disappeared.

Perhaps this sounds a bit dubious because some sorts of pain resolve quickly on their own (although not typically the misery caused by ant bites and stings). But here's an important consideration that further substantiates the validity of a temporal control, at least in some cases. This is essentially the same sort of reasoning we applied in the case of our negative control for the effectiveness of the new drug. By not treating some of the patients, we attempt to show that had we not given the drug to the treatment group, its members would show no signs of improvement, because the only difference between them and the negative control was the treatment. We try to test, by using a negative control, a certain counterfactual about the difference between an initial state of a patient (before the drug) and the new state of the patient (after the drug). In our earlier example, while we can only reason counterfactually about the state of the researchers' pain before and after drinking the tea, we can mimic the before-and-after difference in the clinical trial by using patients to whom we don't give the drug. Of course, we might've done the same thing in the Amazonian village, but the ant bites were agonizing and nobody would've wanted to take the chance of suffering while the others happily reported relief. The ethical issues associated with the use of controls (e.g., is it right to let some people in a control group hemorrhage horrifically when testing the effectiveness of a treatment for Ebola?) will be considered in a later section. Even without the ethical problems, the relation between controls and counterfactual reasoning might be difficult to grasp. So, let's clarify it and then return to the role of controls in experimentation.

6.2.1 Counterfactuals

Consider the following statement: (a) "If Bob has a headache, he takes an aspirin." Statement (a) is called a *conditional*, and the if-part is called the *antecedent*, while the then-part is called the *consequent*. Counterfactuals are simply conditionals the antecedent of which is (taken to be) false: (b) "If Bob were having a headache (but he isn't), he would take an aspirin." We usually don't mention the stuff in parentheses, because it is clear from the context of the conversation that the antecedent is false. If you knew that Bob *is* in fact suffering a headache, it would be odd for you to say "If Bob were having a headache (and he is)…"

We use counterfactual conditionals, or counterfactuals for short, all the time. "If Joe hadn't had those five shots, he wouldn't be drunk." "If I had not been speeding, I wouldn't have rolled the car." "If only I had not spent all my money on new clothes, I would have enough to pay my rent." As the last two examples show, counterfactuals are the currency of regret. As the first example shows, they are also, on some views at least, the currency of causation which interest scientists. The last section revealed that counterfactual reasoning is also involved in temporal controls. We compare the experience of relief with the pain felt before the tea was consumed and conclude that had the scientists not drunk the tea, they'd continue to show the same characteristics that they did before the treatment. But how do we know this?

6.2.2 Possible Worlds

The key here is to think about *possibilities*. Go back to Bob and statement (b): "If Bob were having a headache (but he isn't), he would take an aspirin." Clearly, we assume in expressing (b) that it is possible for Bob to be having a headache, and we also assume that it is possible that he takes an aspirin. The fact that we use the subjunctive mood in (b) simply indicates that while it is possible for Bob to be in pain and therefore to take an aspirin, it is not *actually* the case. There is a fancy way of describing this. If it is possible for Bob to have a headache, then there is a *possible world* in which he is in fact pained. This possible world is, however, not the *actual world*, for in actuality, he is not in pain. There is also a possible world in which he takes an aspirin, which is, again, not the actual world. So when you say "If Bob were having a headache, he would take an aspirin," you are saying that in the possible world in which Bob is in pain, he takes an aspirin. In fact, you seem to be saying something stronger; you seem to be saying that in *all* the possible worlds in which Bob has a headache, he takes an aspirin.

The widespread use of counterfactuals – ordinary regret, causal claims, and temporal controls – is sufficient to cast doubt on a specific form of the doctrine of *empiricism*. According to this doctrine, knowledge about the external world is exclusively based on observation, or, which comes to the same thing, empirical data. However, we can't be too strict about this. Otherwise, we could not have any causal knowledge or knowledge about temporal controls. According to at least one highly influential theory of causation, whenever we reason about causation, we are engaged in counterfactual reasoning, which involves considering what would be true in other possible worlds (we will explore other approaches to causation in Chapter 10). Clearly, we do not have empirical access to those possible worlds. We cannot literally observe what would happen, if something that in fact occurred, had not occurred. We can't look into other possible worlds, but if it is the truth of statements *there* that determines whether or not, for example, a causal claim is true *here*, then strict empiricism, the doctrine that *all* knowledge rests exclusively on data (observation), is not viable.

6.2.3 Counterfactuals and Controls

Our discussion of counterfactuals raises a very difficult question, namely: How do we know what is or is not true in other possible worlds? In its generic form, this question has produced an immense literature, which, however, we can ignore for our purposes. Here, we are just interested in the relation between controls and counterfactuals. The idea of connecting the two has surfaced, e.g., in the manipulationist approach to causation (see Chapter 10). In his book, *Making Things Happen*, American philosopher James Woodward states that in order to evaluate the counterfactual that if some patients had been given a drug, they would have recovered, we can't give and not give the drug to the same individual. Instead, we "employ a more indirect method: we divide the subjects with the disease into two groups, one that receives the drug and the other (the control group) that does not, and then observe the incidence of recovery in the two groups." (p. 95) In other words, with such controls, we try to answer the question whether it is true that if the patient had received the treatment, she would have gotten better. And we try to answer this question by bringing about a situation in which there is a patient who gets the treatment and then observe whether or not she is getting better, as compared to someone who didn't get the drug.

Highly simplified, that process can also be described as follows. We are interested in knowing whether Jane would have gotten better without the treatment. So we select Sue to see whether she gets better with no treatment. If she doesn't, we conclude, roughly speaking, that the drug made a difference for Jane. This inference is of course only legitimate if we can make sure that Sue can actually stand in for Jane in the sense that we can learn from observing her what would have happened to Jane, had she not received the treatment. This means that Sue better be like Jane in the relevant respects. Obviously, the two are not the same in all respects. But as long as we can be reasonably certain that they are the same with respect to the features that matter for the comparison, things should be okay – not certain by any means, but acceptable.

What are those features that matter for the comparison? Obviously, both Sue and Jane must have the same condition, be it a headache, Ebola, or what have you. You wouldn't want to see whether a new drug helps with recurring headaches by giving it to Jane but not to Sue, if Sue doesn't have recurring headaches.

Notice that in using a negative control, we assume that Sue doesn't have any features that make her special and would explain why she got better without treatment (for example), despite the fact that Jane wouldn't have gotten better. But what's the justification for this assumption? This is the place where *randomization* becomes crucial.

6.3 Randomization

A central tenet of experimentation is the necessity for randomizing the assignment of treatments to subjects (whether they are individual organisms, plots of land, or test tubes). But what is the purpose of randomization; what do we seek to achieve with this approach? There are two answers to this question.

6.3.1 Bias

First, we worry about the possibility of bias. If the scientist developing the pain medication stood to become rich if the drug was successful, then she might choose individuals for the treatment group who had characteristics making them more likely to respond favorably (e.g., they seem to have a healthier constitution). Or conversely, she might select individuals for the control group who appeared more likely to have persistent headaches (e.g., they answered questions prior to the experiment indicating that their headaches lasted for many days), so that the rate of spontaneous pain remission would be unrealistically low making the medication appear more effective. In such cases, we'd say that the scientist was acting with explicit bias.

Such instances of shaping the treatment or control groups to intentionally favor a desired outcome are rare in research – or at least we hope so. The greater concern is that of implicit bias in which a scientist unconsciously chooses subjects for treatment with particular features. For example, imagine that the pain medication is being tested on rats. The researcher sends her technician to the rat colony with orders to bring back a batch of animals which will receive the treatment and then, while the drug is being administered, to retrieve another batch of subjects to serve as the controls. The technician goes into the room with the laboratory animals and, being anxious about the possibility of being bitten, understandably selects rats that seem less aggressive. These become the treatment subjects, while somewhat more agitated rats selected subsequently go into the control group. It's not difficult to imagine that the behavioral dispositions of the animals could confound the effects of the pain medication. Perhaps more placid animals are less likely to exhibit symptoms of pain or maybe more aggressive animals have greater resistance to pain. This is just one way that an implicit bias can confound experimental results. Even something as seemingly innocuous as choosing test tubes could become a problem if, for example, the scientist inadvertently selected the cleanest glassware first and allotted these tubes to the treatment group. Returning to our earlier examples of controls, consider:

(1) An entomologist wants to determine if a new chemical is toxic to insects. In selecting experimental subjects, he reaches into the cockroach colony and grabs the first insects he can capture for the treatment group. In so doing, he inadvertently selects subjects that are healthier and larger than a later group of controls. As such, he underestimates the effectiveness of the novel insecticide.

(2) A psychologist wants to determine if "priming" people with an image of a baby would make them more likely to act kindly. The scientist chose older subjects for the treatment, based on an implicit bias that they'd be more likely to have experience with babies and a greater inclination to be generous to panhandlers in the subsequent test. As such, the results overestimated the effect of "priming" because the treatment group was particularly empathetic.

In all of these cases, the problem of explicit or implicit bias can be solved by randomly assigning the treatment and control to the experimental subjects. For the purposes of experimentalists, randomness can be understood as *equiprobability*. That is, each individual subject (whether an organism, field of corn, or test tube) has the same chance of being selected. More specifically, what the elimination of bias requires is *subjective equiprobability*, which means that from the researcher's perspective, each subject is equally likely to be chosen. As such, randomization could be achieved by the use of a table of random numbers or similar mechanism (e.g., as each rat, cockroach, moth, or human enters the study, if the next random number in the table is even the subject receives the treatment, if odd the individual receives the control). But what's important is that the researcher does not allow bias, so she could flip a coin, ask a technician to pick a number between one and ten, or otherwise have no influence over the assignment of treatments.

6.3.2 Unnoticed but Relevant Differences

The second worry involves the possibility of systematic differences that don't arise from experimenter bias. Let's suppose that a scientist wants to test a fertilizer on the yield of Brussels sprouts. He finds a suitable field site and assigns treatments to the rows closest to the road so the application equipment doesn't need to be dragged through the field and upset the grower. The scientist has no sense that these rows are any more or less likely to respond to the treatment; the choice is purely a matter of benign convenience. However, it turns out that plants closest to the road get runoff water and the dust from passing vehicles contains nutrients that have otherwise been lost from the field after many years of planting and harvesting. Without randomizing the choice of the rows receiving the fertilizer, the scientist unwittingly overestimates the effectiveness of the fertilizer on Brussels sprouts yields.

This example of a fertilizer study raises two final issues concerning bias and randomization. After the scientist has applied the fertilizer, he'll need to take samples from the field rather than counting and weighing every single Brussels sprout in the treatment and control plots. This means another source of potential bias is introduced. As with the assignment of treatment and control, the researcher might introduce explicit or implicit bias, as well as systematic differences. That is, he might pick the biggest Brussels sprouts from the fertilized plots since his new-fangled product will make him rich if it works, or he might simply pick the topmost sprouts which could end up being nonrepresentative of the yield, or he might select sprouts from the first dozen plants in each row because his back gets tired from stooping. Again, randomization is an effective means to the ends of avoiding all these problems.

Another consideration is the practical constraints on randomization. Under optimal conditions, a researcher randomizes at every step in which bias or systematic differences might confound the results. But in reality, this is rarely done. Usually, a kind of haphazardness serves the purpose as the researcher simply nabs rats or test tubes in an arbitrary manner. Other times, cost considerations make randomization difficult. Imagine that a researcher is testing a new insecticide for controlling grasshoppers on very large grassland plots, each being several hundred acres. In addition, the research protocol requires sampling daily in the treated plots but weekly in the control plots. Finally, consider that the prevailing winds are from west to east. If the treatments are randomly assigned to plots, the aerial applicator might have to fly long distances to reach each plot – and this inefficiency could be prohibitively expensive. Also, the technicians sampling the treated plots will spend many more hours walking from one to another than if the plots were adjacent to one another. And, if sampling is done randomly within each plot, then drift from the treated plots into the edge of the untreated plots might reduce insect densities below that which would be representative of a true control. All of these factors will be weighed by the researcher and the closest approximation to full randomization will be used within the logistical constraints of the study. If the scientist then accurately reports how the treatments were assigned and the sampling was conducted, the scientific community can decide whether the methods compromised the integrity of the experiment.

However, apart from practical problems confronting randomization, there is also a deep theoretical problem, which can be seen by relating randomization to confounds more explicitly. Confounds, such as bias and unnoticed differences, are factors that might skew the results of the treatment. For example, when we investigate the efficacy of a treatment for cardiovascular disease, age might be such a confounding factor in that the drug might be less efficacious in an older population. Randomization, from this perspective, has the purpose of achieving a balance of potential or known confounds across treatment and control groups. In our example, it would be a bad idea to have only young people in the treatment group, and only elderly in the control. We should randomize so that age is balanced between treatment and control.

Now here is the problem. Even if it is true that randomization balances each one of the known or potentially confounding factors, it doesn't follow that it balances all of them. To think otherwise is to commit an error in probabilistic reasoning. Suppose the probability of achieving a balance for each factor is 0.9. If we assume that there are 10 known or potential confounds, the probability of achieving a balance for all of them is the *product* of the individual balancing probabilities. This product is below 0.4, a much lower probability than the probability for each possible confound taken individually. Thus, in most actual trials, which are characterized by a plethora of potential confounds, randomization does a bad job in balancing all of the confounds. There are attempts in the recent literature on confounds to deal with this problem by, e.g., replacing the ideal of achieving near perfect balance of confounds with something more realistic. The details, however, would lead us into technical discussions far beyond the scope of this book, however fascinating they are.

As an alternative to randomization, we could take into account the different known confounds through a technique called *stratification*. This is the process of grouping together those subjects that share a common confound, such as age group or sex. But this gives rise to the need of larger samples to represent the strata, and thus, if we introduce even a handful, the sample size required becomes a logistic nightmare.

6.3.3 Ethical Concerns

Nevertheless, randomization is important for ruling out confounds. The members of the control group must not be systematically different from the members of the treatment group so that we can be reasonably sure that the only relevant difference is the treatment. Randomization, by minimizing bias and unnoticed but relevant differences, is, within the limits discussed, a good tool for achieving this goal. However, there also may be ethical concerns which have to be taken into account when deciding on a proper control regimen.

Let us briefly consider three ethical issues that can arise with the use of controls. These concerns are particularly relevant to medical trials, but it is not difficult to imagine how these issues could be manifest in the development of anything that is intended to benefit sentient beings.

First, when a drug is given in real life, the patient knows that the medicine has met the standards of effectiveness, so the individual is implicitly told, "This is a real drug." But in clinical trials, subjects are told that they may receive the actual drug or a placebo – and this leads to rather different expectations than in actual practice. The solution is to use a "balanced placebo" approach so that in addition to the negative control group, there are two placebo and two treatment groups with half of each group being told they are receiving a placebo and half being told they are getting the real deal. The problem here is that the researcher is lying to half of each group, thus violating the principle of informed consent. A related problem is that in the real world, doctors give medicines with the implicit (and usually explicit) message that the drug will work. In a sense, if we want patients to get better, then shouldn't we pursue medicines that are most effective in combination with the patient's expectations through their doctor's statements (i.e., a drug that works best when a doctor states that she expects the patient to improve by taking the medicine)?

The second ethical issue arises with the standard called *clinical equipoise*. In short, the researcher must believe that the treatment and the control (positive and negative) groups are getting functionally equal care. So, when a standard therapy is available, it ought to be the case that everyone in the study is receiving medical care at least equal to that treatment. Obviously, if some sick people are getting a placebo, which the researcher fully knows, then the ethical standard is violated. One way to circumvent this problem is the use of active placebos, if they are available. This strategy tests a new drug's efficacy against a drug that is already used for treating a certain disease. Instead of an inert substance, the control group receives the drug already available, thus lowering the potential risk of adverse outcomes.

The final worry arises when, in the course of a clinical trial for either a mild or serious condition, the treatment group shows dramatic side-effects or improvement. In the former case, the research might be terminated based on the unexpected harm not justifying the completion of the study given the mild illness. In the latter case, the trial might be ended early to allow the control group (and others) access to a demonstrably effective treatment for a serious disorder that would otherwise be withheld for some time. The challenge is how much of a "clinically important difference" is necessary to ethically require that the control group be provided with the treatment.

6.4 Conclusion

Having analyzed the nature of treatments and controls, along with the vital role of counterfactual reasoning in drawing conclusions, our task in the next chapter is to apply these concepts to the *sine qua non* of science, experiment. There is no more characteristic process in science than the experimentation. But vitally important conceptual issues arise when developing and learning from experiments. And in terms of the big picture of science, we must rigorously analyze how such explorations into the workings of the natural world relate to the development of theories.

Annotated Bibliography

Jonathan Fuller, 2019, "The Confounding Question of Confounding Causes in Randomized Trials," in *The British Journal for the Philosophy of Science* 70(3): 901–926.

Argues that a complete balancing of confounders between treatment and control groups is an unachievable but also unnecessary ideal that should be replaced.

Ian Hacking, 1988, "Telepathy: Origins of Randomization in Experimental Design," in *Isis* 79: 427–451.

A fascinating account of the widespread interested in psychic phenomena during the nineteenth century and how it eventually lead to an emphasis on randomization and controls.

Alfredo Morabia, 2011, "History of the Modern Epidemiological Concept of Confounding," in *Journal of Epidemiology and Community Health* 65: 297–300.

Available at https://jech.bmj.com/content/jech/65/4/297.full.pdf Starting with the issue of group noncomparability, it follows the development of one particular concept of confounds to the modern conception that is based on potential outcome contrasts.

7

PHYSICAL EXPERIMENTS AND THEIR DESIGN

7.1 Historical Remarks

A distinctive feature of contemporary science is its reliance on experiments. Considering the entire documented history of our attempts to understand the world, which began at around 600 BCE (more than 2,500 years ago), this is a relatively new phenomenon. Experimental results started to inform the development of empirical theories in a systematic way no earlier than the seventeenth century, less than 400 years ago. The great pioneer of science, Aristotle (384–322 BCE), never gave experiments a second thought. For example, Aristotle viewed physics as being concerned with the principles of natural change, including change of place (motion), change in maturity (growing), and change in knowledge (learning). The Greek term "physein," from which our term "physics" derives, denotes such natural change. Interfering with what happens naturally, through experiments, would have been deemed inimical to the goals of the physicist.

This negative attitude toward experiment, as an undue interference with natural processes, remained dominant until the seventeenth century, when Francis Bacon (1561–1626; he's the guy who coined the phrase "knowledge is power") apparently became tired of merely passively observing and started to advocate a more interactive approach. Using an infamous simile, Bacon suggested that the scientist should "torture nature" in order to discover the secrets of the world around us. The expression is harsh, but there remains a sense that our understanding of the natural world requires experimental

This is Philosophy of Science: An Introduction, First Edition. Franz-Peter Griesmaier and Jeffrey A. Lockwood.
© 2022 John Wiley & Sons, Inc. Published 2022 by John Wiley & Sons, Inc.

manipulations. However, it wasn't until the seventeenth century that experiments became an integral part of science. In his *Opticks* (1704), Isaac Newton (1643–1727), for example, describes in great detail how he determined experimentally that white light is composed of colored light. Maybe the greatest number of early systematic experiments were those carried out by Robert Boyle (1627–1691) with vacuum pumps.

These very brief remarks on the history of science highlight a distinction that is still routinely made: the difference between experiments and mere observations. The remarks also suggest that the distinction turns on the notion of *interference*, which seems to be present in experimentation only, but not in observation (at least when conducted in a careful manner – a very challenging matter for animal behaviorists and cultural anthropologists, for example). After a discussion of some of the issues pertaining to the design of experiments, we turn our attention to what can be learned from experimental results.

7.2 Setting Experimental Parameters

Even before an experiment begins, a researcher must decide on background or baseline conditions. Let's assume that a scientist wanted to determine if a commercial chemical advertised as a repellent for gardeners trying to protect their plants would repel grasshoppers and prevent feeding damage. The design might consist of putting individual insects in a small cage holding a spinach plant at the two-leaf stage with one leaf treated with the putative repellent and one left untreated as a control. The entomologist might have decided to apply various doses of the chemical, perhaps half the label-recommended rate, the recommended rate, and twice that rate. And perhaps he determined, based on experience with the variability in grasshopper feeding behavior, to conduct 10 replicates of each treatment. Fine, but there are at least two parameters missing from this design.

First, grasshoppers taken from a laboratory colony where food is abundant won't be very hungry, so they won't rush to chew on either leaf of the test plant. Should the experimenter wait an hour, 12 or 24 or even longer to be sure that the subjects will actually be making a decision between the two leaves? The entomologist might remove grasshoppers from the colony and cage them without food for 6, 12, 24, and 48 hours (based on general background knowledge of insect digestion rates), then place them with untreated plants to see how rapidly they begin feeding.

Second, how long should the experimenter wait before assessing the amount of feeding? If too little time passes, then only insubstantial nibbling will occur, but if too much time passes, both leaves might be entirely consumed as the grasshopper is hungry enough to even munch on the otherwise repellent option. To set this parameter, the researcher might take hungry grasshoppers (it turns out that 24 hours without food is about right), put them on untreated plants and assess leaf damage at 3-hour intervals to determine the time at which there is significant feeding but not the consumption of both leaves.

And so, prior to the designed experiment, it is often necessary to conduct systematic observations to set the baseline parameters in a way that allows a scientist to reliably assess the effect of the treatment. This takes us to the next, crucial question: What variable(s) will be assessed in the experiment?

7.3 Dependent and Independent Variables

When conducting an experiment, scientists differentiate between the independent and dependent variable. The independent variable is systematically changed (e.g., the dose of a drug) or naturally fixed (e.g., time), while the experimenter monitors the dependent variable (e.g., pain) to determine the effects of the independent variable. For example, does pain (at some given time) diminish with increasing dose, or does pain (at some fixed dose) diminish steadily over time?

The problem of bias emerges again, and scientists have devised approaches to avoid the associated pitfalls. The "gold standard" is called a double-blind experiment, in which neither the experimenters (individuals administering the treatment and monitoring the results) nor the subject receiving the treatment knows whether the substance is actually the material being tested or a positive control (placebo). Such designs make sense in clinical trials of drugs in which expectations can influence the results. However, in experiments involving field plots, or animal models, or nonliving systems (chemical reactions or physical phenomena), the subjects are "blind" by virtue of their lack of expectations – gardens, grasshoppers, and galaxies aren't concerned with a scientist's interests. That said, the experimenter might still exhibit bias when administering the treatment or assessing results. The scientist testing his novel fertilizer on Brussels sprouts might make the applications more carefully to rows of plants if he knew the tank contained fertilizer rather than an inert control materials. Or in assessing the material, he might unintentionally seek out the healthiest plants to harvest if he knew a plot had been sprayed with the fertilizer. In this case and similar ones, randomization serves to effectively "blind" the researcher. While randomizing the rows receiving fertilizer would avoid potential bias during the application phase of the experiment, doing so might not allow the researcher to optimize the analysis of factors influencing plant growth. What if fields have greater soil moisture in their lower areas or greater warmth on their south-facing regions?

7.3.1 Stratifying to Isolate Relevant Factors

Recall from last chapter that researchers are not restricted to randomizing across the entirety of the potential subjects. If the scientist has reason to expect that certain qualities of the subjects or experimental setting will play an important role in the effect under consideration (e.g., soil moisture or solar radiation), we can stratify our sampling based on these factors. That is, the experiment can be designed so that extraneous factors are systematically included, thereby allowing the subsequent analysis to take into account – and eliminate the influence of – these confounding elements.

For example, let's assume a psychologist has found that "priming" people with the image of a human baby makes them more likely to give money to a panhandler. Now the scientist wants to determine whether the duration of seeing the photograph matters. So she decides to test teenagers, adults, and seniors. However, her assistants noted a general tendency of people to be less generous to the panhandler on Mondays (when people are more grumpy) and more generous on Fridays (when people are less stressed). A nonstratified way to run the experiment would be to randomly apply the treatments on any given day. However, the researcher could stratify the experiment by testing each of the age groups on each day of the week. The only randomization would be choosing which subject in each age group received which exposure on each day.

One of the serious challenges facing an experimentalist is that of deciding which of these countless factors should be stratified, since doing so imposes logistical constraints. In our example, there was reason to think that day of the week mattered. But how about the gender, weight, or zodiac sign of the subject? Or whether the subject possesses more or less than $20 when approached? What constitutes a factor that is sufficiently relevant to warrant stratifying versus one that can be reasonably randomized? To identify potential confounds, the scientist might turn to the published literature, hunches in light of background beliefs, or to observations about the system under investigation. This sort of question pertains to not only the choice of extraneous factors but to what a researcher takes to be independent variables worth investigating.

7.3.2 Determining Relevance

The above considerations point to a deep philosophical problem for practicing scientists, namely, how to determine relevance. Maybe the closest we have to a definition of relevance applicable to science can be found in discussions of statistical relevance. The idea is fairly simple. An event A is statistically relevant to some event B if and only if it is the case that the probability of B, conditional on A, is different from the probability of B simpliciter. For example, the probability that a die comes up 5 after having been cast is 0.17 simpliciter, but is 0.33 conditional on the information that the die shows an odd number.

Unfortunately, statistical relevance is of no help in trying to <u>find</u> a relevant factor during experimental design. Once I have identified a potentially relevant factor, I can use the statistical relevance test to confirm that the factor is indeed relevant. But I can't use statistical information to identify a potentially relevant factor. It might seem that I could, in principle, apply the statistical relevance test to all available factors, and thus find the actually relevant ones. But that's clearly not realistic, and it would be a waste of time: Should we really test whether or not the color of the underwear of the experimenter makes a difference to the probabilities in an experiment on igneous rocks?

What scientists do is rely on background theories and beliefs to help them cut down on the number of potentially relevant factors. Colors of underwear, such theories tell us, won't influence the melting point of igneous rocks, but pressure might. Of course, this leads to the further problem of determining the relevance of background beliefs to the issue at hand. Here, scientists seem to be guided by some rough and ready categorizations. For example, questions of fashion are unrelated to questions of geology. Of course, we might be quite wrong in our categorizations, as they are in large part informed by the theories of the world we currently accept, and those theories might be radically wrong. As the philosopher and historian of science, Thomas Kuhn, has famously argued, what factors are deemed to be relevant to a given phenomenon might undergo radical shifts during scientific revolutions. Maybe this means that there is no ahistorical concept of relevance. But we'll have to leave it at that.

7.4 Learning from Experiment

Why do we carry out experiments? Clearly, because we can learn something, but about what? The most obvious answer is also the least interesting: We can learn something about how the things *actually used* in the experiment interact, how the dependent variable changes as a result of various interventions, or whether there is any kind of regularity if the experiment is repeated. But in the majority of cases, we are not really interested solely in the features of the things (events, processes, etc.) used in the experiment. Rather, we are interested in learning, *by* using them, something about similar things in the world. For example, if we test a drug for side effects on model animals, we hope to learn thereby something about potential side effects on humans. Or, if we measure the half-life of a particular chemical, we hope to learn what the half-life is of all the chemicals of the same kind.

Obviously, one important use of experiments is to learn about a whole class of things to which those used in the experiment are relevantly similar. In this sense, experiments constitute models, the manipulation of which hopefully allows us to draw inferences about the things they represent which are not experimented upon. We will talk about the relation between models and real-life system in Chapter 9. For now, we want to concentrate on the process of learning from experiments, seen as models, and in particular, on the process of *extrapolation* from the experiments to other similar systems.

One of the main problems with extrapolation is that what we learn from the experiment might be idiosyncratic to the things actually used. In Chapter 1, we pointed out that various sampling procedures increase the probability that the sample is representative of the population. We can think of experiments as a sort of sample: Make a model of the essential features of a whole class of real-life systems, manipulate it in the laboratory or field, record the outcomes, and make appropriate extrapolations. Determining the essential features themselves is tricky. We usually start with an educated guess, guided by some general background assumptions similar to those involved in decisions about relevance, but always ready to change our view on essential features in light of empirical evidence. Replicability and various techniques involved in evaluating the data can perhaps partly be seen as reactions to the worry about essential features.

7.4.1 Replication: How to Be Confident

The core principle that allows scientists to express confidence in their findings as representative is replication – in short, repeatedly finding the same results. Replication can be sought both within and among experiments. In principle, one can conduct an experiment on a single individual if time is used as a control. I can rub hair tonic on my balding scalp and observe whether, over time, my hair loss declines (e.g., counting the number of hairs in my brush). A more plausible test would be to involve my brother, who shares many of my genes (including, it appears, the one for hair loss) and apply the tonic to one of us and nothing to the other (a negative control) or perhaps the alcohol which serves as a solvent for the supposedly active ingredient (a positive control). But what if I find that the tonic-treated sibling had just a small decline in the number of hairs caught in his brush? Would I be willing to declare that the tonic prevented, or at least delayed, balding?

No – there might be chance differences in the rate of hair loss between us. What I'd seek is a number of test subjects or replicates. So, I round up a couple dozen of my 50-something friends, randomize the treatments, and measure hair loss. If I see similar results across many individuals, I gain confidence in my conclusions. Suppose I find a pattern of hair retention in the treatment group relative to the control, would the scientific community accept that I've found a cure to baldness?

Again, no. There would be suspicion that my single study was unrepresentative because only a small group of my buddies was involved. Maybe the response was particular to a tiny subsample of humanity – perhaps the diets of the tested group confounded the results, or maybe it could be that younger or older men respond differently, or maybe men in Japan, Brazil, or Sweden would respond differently, and what about women (they experience hair loss as well)? Thus, replication has to be accompanied by appropriate randomization procedures to minimize the risk of extrapolating from unrepresentative samples.

There is another concern regarding the nature of replication – the problem of so-called *pseudoreplication*. True replicates should be independent from one another, so that the response of each is less likely to depend on confounding factors. Imagine that in the experiment with the balding buddies, I measured hair loss in the dozen treatment and control subjects every 2 weeks for a full year. That would generate $12 \times 26 = 312$ data points per group – not a bad sample size for statistical analysis. The problem is that if some fellow in the treatment group was genetically disposed to retain hair, his response would have been counted 26 times. This would be very different than if the experiment had actually involved 312 individuals in each group. The results of repeated measurements indicate a much larger sample size and greater statistical power than is justified. Valid replicates should not be dependent on one another – and in our example, the fellow's hair retention will be highly correlated with his retention 2 weeks ago, and that from 2 weeks earlier, etc.

Just what constitutes sufficient grounds for independence of replicates is not always entirely clear. Thinking back to the Brussels sprouts experiment, it might be the case that rows of plants separated by a few feet share the same confounds, such as being close to a road. But does valid replication mean applying the treatment to different fields separated by hundreds of feet, or entirely different farms in the same valley, or completely separate regions? The answer lies largely in the scope of one's conclusions. If all of the Brussels sprouts replicates were in the same agricultural district, then drawing a conclusion about the fertilizer's effectiveness in this region would be justified. But extrapolating beyond the limits of the replicates would be problematical, particularly if there was reason to believe that various agricultural districts growing Brussels sprouts have different soil types, pest problems, etc. And again, even if the conclusions are limited to the spatial context in which the experiment was conducted, we must be careful in extrapolating beyond the year in which the research took place as idiosyncratic weather conditions might have contributed to the apparent success (or failure) of the treatment. Hence, replication in time – as well as space – may be important to making a scientific claim (this would not be the case for most experiments involving subjects that aren't likely to change with time, such as protons, molecules, and standard strains of laboratory animals).

Although replication is a desirable means for justifying scientific claims, both within- and between-experiment replication is limited in practice. Thousands of replicates of most experimental subjects is cost prohibitive and inefficient. More importantly, some experiments might involve subjects that are simply too large to replicate in any meaningful way. There's only one Earth, so we can't replicate planetary-scale treatments, such as adding particular chemicals to the atmosphere. Likewise, if we think of the 1988 Yellowstone fires as a kind of treatment, there is no way to replicate such an enormous swath of land. As for replicable experiments, the emphasis in scientific funding and publication is on gaining new knowledge, not confirming past findings. In the course of conducting new experiments, one might begin with replicating the foundational work of others – and, as has been recently seen with systematic efforts to repeat psychology experiments, an estimated 36% – 47% of published studies may not be replicable.[1] And even within a single experiment some results might be suspicious in context of the rest of the data which leads to the next analytical tactic.

7.4.2 Misleading Evidence: Cleaning up the Data

In the course of an experiment, it is common to have one or more results that differ drastically from the general trend. Including this data in the analysis increases the variation and may obscure actual differences between the treatment(s) and control(s). Such data is referred to as being an *outlier* because it lies outside of what the scientist takes to be representative or valid results. There are three reasons that such data arises – but there is no simple, universal approach to objectively determine which explanation is correct and what should be done.

First, the anomalous data could indicate some important phenomenon. Returning to our hair tonic experiment, suppose the table of data shows that one of the balding subjects grew an extraordinarily full head of hair in just 2 weeks, while everyone else exhibited very modest increases over the course of several months. This outlier could indicate that something about that individual (genetics, diet, etc.) produced truly remarkable results – and just what this feature was could warrant further investigation.

Second, in any experiment involving a large number of subjects, the statistical distribution of responses will include some data that are far from the central tendency of the group. We wouldn't want to exclude results that are at the upper or lower bounds of a distribution just because they fail to accord with the average response. There are, however, various analytical methods that a researcher can use to determine whether a seemingly anomalous data point or several such points are beyond what would be expected given the underlying statistical distribution (e.g., researchers often assume a normal distribution and are suspicious of data that fall outside of three standard deviations – a measure of the expected variance or dispersion). This leads us to the third explanation for outliers.

Sometimes things go wrong and a result reflects not the way the world works but the way in which experiments are prone to human error. The data point indicating the bushy-haired response to tonic might have been a transcription error (e.g., 130 hairs per cm^2 was accidently entered as 130 hairs per mm^2), or the hair-counting technician was untrained, careless, inebriated, or deceitful.

From a philosophical perspective, what these ways of treating misleading evidence show is that we cannot *just* rely on our experimental data if we want to understand the relevant real-life systems. Sometimes the data do not reflect only the phenomenon we are interested in, but also features of, and possible problems with, the process of data collection and recording.

Sometimes, we use general knowledge about statistical distributions for cleaning up the data. In those cases, we have theoretical expectations about what the data should look like – our expectations trump (some of) the outcomes of our experiments. Finally, there may be genuinely new phenomena revealed by our data.

What this shows is the existence of a feedback loop in the development of theories. We start with data and use background theories to evaluate and clean up those data. But we can also use the data to change our theories. In other words, we don't just collect data, then conjecture theories on their basis, and finally test those theories by producing new data in new experiments. This linear path is an egregious oversimplification of the actual path from data to theory, which often, if not always, involves considerations of which data we should take into account. And there is no general rule as to when a researcher throws out extraordinary data to preserve a theory and when a theory is abandoned or modified to accommodate unexpected data. Nor is there a single, best method for interpreting whatever data are deemed credible.

7.4.3 Data Reduction and Curve Fitting: Promises and Pitfalls

There are several approaches to taking the "raw" data from an experiment and transforming it into interpretable results. One set of methods involves an expression of central tendency – the most common version being the mean or average. This statistic represents the overall feature of the data, as long as there is a normal distribution with few outliers. Suppose a social scientist wanted to express the incomes of physics departments with a single measure. She collects the salary data and reports that the average physics faculty member earns $251,000. The scientists are quite amazed as nobody earns even half this amount – except the Nobel Prize laureate in one institution. In fact, the salaries are (in thousands), 77, 79, 79, 81, 84, 90, 96, 97, 104, 109, 116, 123, and 2,132. So the mean, excluding the one guy the administration pays extremely well for the prestige he brings to the university, is $95,000.

The sociologist could consider the Nobel laureate an outlier and omit that data point, or she could use the median income – the salary above which half earn more and half earn less, which is $96,000. This measure of central tendency is not heavily influenced by outlying or extreme data. Another option would be to use the mode, or the most common value in the data set. This can be informative, particularly for large amounts of data, but in our case the value would be $79,000 which is not representative. However, the modal value for testicles is two (more than half of the human population being men and most having a pair of gonads), which is a fine expression of this feature in our species. Conversely, the mean would be ~ 1 which would mischaracterize humankind, because very few people have a single testicle. When the data are distributed with perfect normality (the classic bell curve), the mean, median, and mode are identical.

The other common approach to summarizing data is a graphical representation which shows the relationship between the dependent and independent variables. But a simple scatterplot doesn't reveal any underlying pattern or curve from which a mathematical equation or model can be derived. How should we construct a curve so that it captures what we believe to be the actual relationship? The simplest method would be to simply connect the data points with a jagged line which provides maximum goodness-of-fit because every point lies on the line. However, this representation is generally unsatisfying because we don't really believe that every data point is measured without error and warrants such an influence on the overall relationship.

Thus, in the 1970s, the Japanese statistician Hirotugu Akaike developed an influential justification for the use of simplicity considerations in curve fitting. In a nutshell, the idea is this. Every data point that we record reflects not only the actual relation between the dependent and independent variables, but also noise, such as friction and air resistance in measuring the period of a pendulum, and measurement error given the limitations of our senses and instruments. These distortions, however, will be different from experiment to experiment. Thus, a curve that fits exactly all data points generated by one experiment is likely to be a bad fit for the data we gather from the next experiment on the same system, or, even more so, on a similar system in a different laboratory. In other words, since maximizing goodness-of-fit for one data set treats noise and error as as part of the actual relation between the dependent and independent variables, the resulting function (i.e., curve) would, in a manner of speaking, generalize noise and error across all systems, as opposed to generalizing just the relation we are after. Simplicity is therefore a way of controlling for noise and error, at least for relatively limited data sets. However, the bigger the data set gets, the less weight should be put on simplicity considerations.

You might ask yourself if there is a way of measuring simplicity, and yes, there is, at least for curves. Simplicity is measured in terms of the so-called adjustable parameters in the algebraic expression of a curve. For example, the simplest curve is a straight line, which is expressed by the equation $y = ax + b$. The letters "a" and "b" are adjustable parameters with "a" being the slope of the line and "b" being the y-intercept. If there is a strict linear relationship between y and x such that for each unit of increase in the independent variable there is a matching unit of increase in the dependent variable, we simply let $a = 1$ and $b = 0$. This results in a curve with a slope of 45 degrees that goes right through the origin of the graph (the 0,0 point). A more complex curve is given by the equation $y = ax^2 + bx + c$. In this case, we have three adjustable parameters. All of those can be freely chosen with an eye toward maximizing goodness-of-fit. If you think about this, it should become clear that the more adjustable parameters we include in our specification of a curve, the easier it becomes to fit the curve exactly to the data. Thus, more adjustable parameters mean greater danger of overfitting (i.e., treating noise and error as actual or informative influences), and therefore a loss of generality. Choosing comparatively simple curves can avoid the error of overfitting the data. There are, however, other errors.

7.5 Types of Errors: Pick Your Poison

In any experiment, the researcher will run the risk of making either a type I or type II error. In the first case, the scientist accepts a *false positive*, concluding that the treatment had some effect on the dependent variable when, in fact, there was no such relationship. In the latter case, the scientist accepts a *false negative*, concluding that the treatment had no effect on the dependent variable when it actually did so. Of course, the scientist would like to make neither kind of mistake, but the two types of error are inversely related – decrease one and the other increases.

The decision as whether one accepts more type I or II errors is oftentimes a matter of the context or the stakes of an experiment or line of research. Suppose that there is a terrible, new disease – or even a pandemic – and a medical researcher is searching desperately for a treatment by screening a small set of prospective compounds. Given the life-or-death stakes, the scientist might rationally decide to accept a larger number of false positives (compounds that appear to be effective but aren't really) so as not to overlook the one drug that might provide a cure. He might figure that further testing will eliminate these errors but at least if there's a viable treatment, it won't be missed. Next imagine that there is a pharmaceutical laboratory attempting to determine if any compound in an enormous collection of substances can chemically eliminate hangnails. Given the low stakes, the researcher might reasonably accept a high rate of false negatives (compounds that appear not to be effective but really are) so as to not waste the company's resources on this venture.

7.6 Relationships between Experiment and Theory

Let's assume that you have carried out some experiments and found some neat regularities. Now, what are you going to do with them? There are theoretical and practical options. The latter include various possibilities of predicting or systematically intervening in and controlling nature, perhaps to our (perceived) advantage. For example, suppose it has been determined that the boiling point of water increases under high pressure, just as the French physicist and mathematician, Denis Papin, realized in the seventeenth century. He applied this knowledge in designing his new *steam digester*, a precursor to both the pressure cooker and the steam engine. Obviously, both of these gadgets proved quite useful, the former for cutting down on cooking times and the latter even more so by ushering in the industrial revolution. For our purposes, however, we will concentrate on the theoretical options for using experimental results.

Experimental results are, from this perspective, just sets of data. In the context of a relevant theory or hypothesis those data become either supporting or undermining evidence for that theory or hypothesis. In the first case, we say that the experiment provides confirming evidence. In the second case, we say that the experiment falsifies, or at least disconfirms the theory. We discussed a number of thorny issues involved in both confirmation and falsification, or disconfirmation, in Chapter 3. We also discussed two further uses of evidence: to *generate* theories and to help us *choose* among competing theories. With respect to theory generation, we pointed out that many of the questions arising in this so-called *context of discovery* are beyond the ken of the philosophy of science seen as a normative discipline. With respect to theory choice, however, it has been proposed that there is a class of experiments that can be used to settle the score in a definitive way: It is sometimes possible to decide which of two incompatible theories is right by using a so-called *crucial experiment*.

7.6.1 *Crucial Experiments*

Perhaps one of the earliest crucial experiments was conducted by Isaac Newton. With it, he tried to establish that white light was composed of elementary colors, as opposed to being simple in itself, as had been standardly believed. The first evidence he had for the claim that white light was so composed resulted from his famous prism experiment. Letting a beam of light go through a glass prism, he was able to produce an elongated spectrum of colors on a screen behind the prism. However, simplifying a bit, his finding that light is a composite was threatened by an alternative hypothesis, namely, that the light had been modified, in a way still to be determined, by the prism.

Thus, there were two competing hypotheses available concerning the effect of the prism on the light beam. Hypothesis one (H_1) was the claim that the prism had separated the light into its elementary component rays. Hypothesis two (H_2) was the claim that the inherently simple light had been merely modified by the prism, much like a stained glass window would. Which hypothesis, H_1 or H_2, should be selected?

Newton's ingenious idea was to set up what is now called a *crucial experiment*. He cut a hole into the screen just wide enough to let one of the colored rays, say, the blue one, pass through, but none of the others. Next, he directed this ray through a second glass prism, identical in kind to the first one. If it were true that the glass prism modified rays of light, it should modify the blue ray as well. If, on the other hand, a prism separates white light into its elements (the component rays), the prism should not have any effect on the blue ray. You may be able to guess the result: No further change in the color of the ray was observed! Thus, H_1 had been shown to be better than H_2. Notice that this doesn't mean that H_1 is true. While H_1 and H_2 could not both be true, both could be false as there could be a third, correct, hypothesis that Newton and others had not considered; two hypotheses that have this feature are called *contrary hypotheses*. Thus, the new experimental evidence favored H_1 and undermined H_2.

Let us isolate the logical form of such a crucial experiment, as it can help us to highlight some of its features and shortcomings. What we do in running a crucial experiment is:

1. Consider a pair of contrary hypotheses, H_1 and H_2.
2. Design an experiment for which H_1 and H_2 give contradictory predictions.
3. Run the experiment and determine which prediction is correct.
4. Conclude that the hypothesis with the correct prediction is favored by this experiment over the other hypothesis.

In some rare cases, we can conclude that the favored hypothesis is in fact true. This happens if the hypotheses we consider are not merely contrary to each other, but in fact are *contradictory hypotheses*. This means that if one is false, the other one *must* be true. In other words, both of them cannot be true, and one of them has to be. In such a case, if the prediction of H_1 turns out true, and the prediction of H_2 false, then, since H_2 is simply the denial of H_1 and has been falsified, H_1 must be true. Or so it seems…until we recall the problems raised by the doctrine of confirmation holism. To see how this works, let's look at a second famous example of an allegedly crucial experiment.

It was conducted by Louis Pasteur in 1859. For centuries, people had thought it possible for life to arise from inanimate matter. Defenders of spontaneous generation posited that fleas could arise from dust and maggots could arise from dead flesh, along with many other commonplace appearances of living organisms. Others maintained the view called univocal generation in which organisms of a particular kind could only be formed via reproduction from progenitors of the same species.

A century before Pasteur, Francesco Redi had shown that maggots don't spontaneously develop from rotting meat. His experiment involved a treatment (jars of meat with gauze over the opening to exclude flies) and control (uncovered jars of meat). As we'd now expect, only the control jars produced maggots – and, to add to the evidence, Redi found maggots on the outside of the gauze, deposited by hopeful flies. But this experiment dealt only with large organisms and a century later the possibility of microbes arising via spontaneous generation was still under debate.

Pasteur took on the case of microbial generation, which represented a prototypical test case of spontaneous generation. Left in the open air, a dish of broth would soon be clouded with microbes that seemingly came from within the inanimate liquid. Pasteur boiled meat broth in a specially designed flask to kill all living organisms in the liquid. The flask had a long, curved neck which prevented airborne material from reaching the broth. The lack of growth in the nutrient-rich medium was taken to be definitive refutation of spontaneous generation – a crucial experiment. The two competing hypotheses were:

H_1: Microbes are generated spontaneously from inanimate matter.

H_2: Microbes are not generated spontaneously from inanimate matter.

As stated, H_1 and H_2 are contradictory because they encompass all of the logical possibilities – there is no third (or further) hypothesis. Thus, a suitable experiment should decide the matter. H_1 entails that if inanimate matter is isolated from the environment, microbes will (still) be generated. H_2 denies this prediction. As reported above, H_2 was clearly favored over H_1 by the experimental outcome. Case closed.

But remember, there are always auxiliary hypotheses that can save a hypothesis (not a third, logical possibility, but a way of explaining away the results favoring H_2). One might argue that the heating of the broth altered the inorganic material such that it was no longer capable of spontaneous generation, hence the only conclusion is that microorganisms don't emerge from this sort of liquid. Or to be more precise, this type of liquid in Pasteur's laboratory; maybe other locales were more conducive to spontaneous generation. And finally, even if one accepted the conclusion that beef broth doesn't give rise to microbial life, this doesn't refute all other inanimate substances as the basis for spontaneous generation.

7.6.2 Are Experiments Theory-Neutral?

Especially in the context of crucial experiments, we often think of the results as being equally acceptable to all parties involved. In other words, we think of evidence as a neutral arbiter (see Chapter 2) that allow us to rationally resolve conflicts of opinion. But is it really the case that experimental results are always equally acceptable to both sides of a dispute? Two considerations put such a view into doubt.

The first issue concerns the possibility that only theories that are embedded in the same wider conceptual framework are truly comparable by how they fit experimental outcomes. In other words, it has been argued that if two theories do not substantially share conceptual foundations, what looks like a convincing experiment to the defenders of one of them looks like irrelevant manipulations of dubious entities to the other camp. We will develop this possibility in detail in Chapter 15.

The second consideration pertains to the need to calibrate the instruments we often use in experiments. Remember the general nature of the calibration problem, which we discussed in Chapter 2. Instruments are introduced to broaden the range of what is observable, or at least measurable. Since these instruments deliver information that was previously inaccessible, or at least not accessible objectively (remember the differences between measuring temperature with your skin versus a thermometer), there is a danger that we might be fooled by the instrument. We might think we are looking at interesting details on the surface of the moon, as Galileo Galilei did, while what may be really going on is that we mistake artifacts produced by the instrument (a telescope in this case) with features of the moon (this is what Galileo's detractors claimed). How can we tell which it is? Galileo solved the problem by pointing his telescope to a little chapel on a hill so far away that the building couldn't be seen with the naked eye, but only with the help of his telescope. He subsequently walked to the hill only to find the same chapel he saw through the telescope. The instrument had been calibrated – at least for relatively short distances. There are, however, cases in which calibration takes on an unmistakable air of circularity. Let's finish with a simple, but obviously contrived, experiment that brings out the danger of circularity lurking behind calibration quite clearly.

A mercury thermometer relies on the principle that substances expand when heated – as the temperature rises, so does the column of mercury. Suppose someone finds this quite incredible and asks for evidence for the principle behind the mercury thermometer. Obviously, one should not test this phenomenon by measuring temperature with a mercury thermometer in order to see whether objects expand when their temperature increases. In such a case, one would be testing the hypothesis by relying on the very assumption under investigation – that the expansion of mercury indicates an increase in temperature, which reflects the principle that heated substances expand.

There may be a way out of the vicious circle in which theory and experiment are interdependent in this case. That is, if there is an instrument that measures the same property using an entirely different principle, then one gains an independent measurement. In our case, a scientist could calibrate the mercury thermometer using a thermometer that relies on a thermocouple (based on particle excitement as a function of temperatures in different metals which generates an electric current) or one using a constant volume of gas (based on increasing pressure which, doesn't entirely avoid the assumption of expansion since this phenomenon is what generates the pressure). However, in other cases, especially those involving sophisticated instrumentation used in very mature disciplines, such as high-energy physics, it is not entirely clear how circularity can be successfully avoided. The problem is similar to that of the theory-ladenness of measurement, which cannot be avoided. We have to learn how to live with it.

7.7 Conclusion

Perhaps many students of science have a sense of what makes for a valid experiment (the "how" of science), however we hope that in this chapter the conceptual foundations for practices such as replication and data reduction have become apparent (the "why" of science). As powerful and efficacious as experiments are to understanding the world, this method is ultimately limited – no approach allows the scientist to avoid mistaken conclusions (either false negatives or positives). Moreover, no experiment is purely objective, untainted by a scientist's beliefs. And as we will see in the next chapter, some important experimental practices don't even appear in laboratory manuals.

Note

1   See Wiggins, B. J., and Christopherson, C. D., "The Replication Crisis in Psychology: An Overview for Theoretical and Philosophical Psychology," Journal of Theoretical and Philosophical Psychology, 2019, Vol. 39, No. 4, pp. 2020–217.

Annotated Bibliography

Malcolm Forster and Elliott Sober, 1994, "How to tell when simpler, more unified, or less ad hoc theories will provide more accurate predictions," in *British Journal for the Philosophy of Science* 45 (1):1–35.

Employs a technical result by statistician Hirotugu Akaike, known as the Akaike Information Criterion, to provide a rationale for using the comparative simplicity of families of curves as a tie-breaker in the context of the curve fitting problem.

Ian Hacking, 1983, *Representing and Intervening*. Cambridge, UK: Cambridge University Press.

One of the few books in the philosophy of science that explicitly address experimentation. The author proposes that one should be realist about those entities that can be manipulated in an experiment, such as electrons.

Deborah Mayo, 1996, *Error and the Growth of Experimental Knowledge*. Chicago and London: The University of Chicago Press.

Arguing against Bayesian approaches to experimentation, Mayo proposes an error-statistical model, according to which experiments are tools for gaining knowledge from error. She argues that a modified Neyman-Pearson statistics is the most promising way to circumvent subjective probabilities often encountered in risk assessment.

8

EXPERIMENTAL METHODS THAT THEY DON'T TEACH

In this chapter, we consider forms of experiments other than the standard version in which a scientist designs and implements a treatment and control to explore some aspect of the natural world. The versions of experiments that we explore are *found, natural,* and *thought* experiments, neither of which conforms to what is often – perhaps mistakenly, as we shall see – taken to be the normal or typical approach to experimentation, namely, experimentation that involves intervening in the physical world. Found experiments involve the physical world but forgo intentional intervention. Thought experiments entail imaginative interventions but do not involve the physical world.

8.1 Found and Natural Experiments

Based on the account of experiments in the previous chapter, one might reasonably think that without an intentional design to systematically vary some feature of the system, it seems that there would be no controls – and hence, no basis for asserting whether or not the treatment had an effect. However, scientific inquiry is not so easily distilled into a simple formula or "how to" manual.

8.1.1 Found Experiments and Unplanned Treatments

Sometimes the world presents the scientist with a set of circumstances that are virtually indistinguishable from what she might have created with

This is Philosophy of Science: An Introduction, First Edition. Franz-Peter Griesmaier and Jeffrey A. Lockwood.
© 2022 John Wiley & Sons, Inc. Published 2022 by John Wiley & Sons, Inc.

a designed experiment. That is to say, sometimes, we get lucky and come across – or we might search for – what could be called a *found experiment*. In such cases, natural processes provide a treatment and a control in a form that arguably constitutes a standard experiment. Let's consider a recent example.

After the great fire in Yellowstone National Park in 1988, some areas were burned and others remained intact, which was a rather typical, mosaic burn pattern for a forest. In an important sense, those two kinds of areas could be seen as treatments and controls, respectively. Suppose I am interested in the effect of vegetation on soil erosion. Looking at burned areas (treatment in this case being the removal of vegetation by fire) and comparing soil erosion (e.g., depth, texture, water-holding capacity) with that of intact areas (unburned control) allows me to come to some conclusions about the relation between vegetation and erosion. Of course, there are bound to be a few confounding factors here. For example, plants in areas with particular kinds of soil might have carried the fire more readily or the heat produced by the wildfire might be an important factor in soil erosion, quite independent from the resulting loss of vegetation. These confounds illustrate a central limitation in found experiments – the treatments are applied in a fortuitous manner. Happenstance alterations of the world might avoid the problem of intentional bias, but relying on chance to determine when and where a treatment takes place is problematical. However, if we characterize experimentation as the use controls to discern the effects of a treatment, not all forms of experimentation involve intentional interference; the scientist can exploit a readymade, natural opportunity.

Our discussion of the Yellowstone fire provides an example of *spatial controls*. We compare a treatment area with an otherwise similar control area. In other cases, we appeal, maybe implicitly, to *temporal controls*. Here, we compare the state of some system (or creature or area, etc.) *after* some significant event with its state *before* that event occurred. In fact, temporal controls are typically used when we make causal claims in cases in which we cannot interfere (for moral reasons, say).

Imagine that a waste incinerator (the treatment) is built near a community of 50,000 and within a couple of years people are asserting there is an unusually high rate of liver cancer. Medical records show that prior to the facility, the community had five cases of liver cancer, reflecting the national average of about 1 case per 10,000 people, and after the incinerator began operations there were 70 cases. Epidemiologists could treat the previous conditions as a temporal control for the claim that the facility is responsible for the change in cancer rates.

To employ the notion of a temporal control in the natural world, consider the Yellowstone fire again. In a sense, the condition before the fire is a control for the treatment (the fire), because we can say that had there been no fire, the higher rate of soil erosion would not have occurred. Such a conclusion would be strengthened by our having extensive records on soil loss from an area of Yellowstone as part of a long-term monitoring program, so that the degree of natural variation would be well known.

So, what would be a conventional experiment analogous to the Yellowstone case? Well, we'd have to start a forest fire! You can see where found experiments can allow us to make inferences about treatments in situations that might otherwise be very problematical. Treatments that cause severe harm might be morally prohibited, and treatments that change conditions over vast areas might be practically impossible (and perhaps legally prohibited). Thus, found experiments play an important role in those areas of scientific research where due to moral or pragmatic considerations, standard experimental practice, and in particular the use of contrasting treatment and control groups, are difficult, if not impossible, to implement. Such research areas may be found in medicine, or in any field in which direct intervention is for all intents and purposes impossible, such as cosmology, paleontology, and much of ecology, to name just a few.

8.1.2 *The Role of Background Theory*

We might well imagine that the world is filled with potential, found experiments. Given how often conditions change, does every event have the capacity to be a found experiment? In one sense, the answer is yes, particularly if we allow that temporal controls are valid. But the challenge for the scientist is to decide which phenomena warrant our attention, which hold the potential to reveal something worthwhile. The answer to this question is largely a matter of the background theory a scientist brings to the world.

Returning to Yellowstone for a moment, imagine that there is a bubbling hot pool with multicolored rings and a few feet away another pool without any coloration whatsoever. A geologist and a microbiologist might respond very differently to these pools. Perhaps the geologist knows that the underlying rocks are uniform in this area, so she's not terribly interested in whether or not the color difference is a function of the geology. Thus she continues on her way. But the microbiologist might wonder whether the bacterial mats responsible for the color represent different biotic communities, while the ecologist might hypothesize microhabitat differences.

What makes for a "good" found experiment are a set of features that lend themselves to meaningful comparisons about phenomena. Different scientific specialists may see the same state of affairs in the world and based on theoretical backgrounds decide that the differences between one time/place and another warrant attention or dismissal. There is no single or simple rule to determine if a change arising from natural events constitutes a found experiment across all theoretical frameworks.

8.1.3 Natural Experiments

Related, but not quite identical, to found experiments are empirical studies in which subjects are exposed to treatment and control conditions according to some factor outside of the experimenter's control. Such occasions often arise when imposing a treatment is difficult or unethical. For example, an epidemiologist might use cell-phone location data to evaluate the chance of infection and death by the COVID-19 in people who had various durations of exposure to individuals testing positive for the virus, or a sociologist might estimate the educational losses by children from schools with various constraints on in-person learning during the pandemic.

8.2 Thought Experiments

We saw in the last chapter that counterfactual reasoning is an essential part of much scientific reasoning. It is connected with our use of controls in experimentation and observation. Now we will encounter another important scientific activity in which counterfactual reasoning plays an important role, viz. the production and use of *thought experiments* (TEs). TEs turn out to be of special significance for some of the most fundamental conceptual innovations in the history of science. It is thus a pity that they receive little or no attention in both scientific textbooks and standard texts in the philosophy of science. Our lengthy discussion of their nature and use is intended to rectify this situation. In addition, they also occupy an intermediary place between physical experimentation and simulation, the latter of which we will briefly discuss below.

8.2.1 Reasoning through Scenarios

Remember one of the central lessons from our discussion of counterfactuals: Some scientific knowledge is not based on observation, because it involves counterfactual reasoning and thus considerations about other possible worlds. During the history of science, there were quite a few instances in which considerations of possible worlds, and what happens there, played a crucial role for either criticizing old theories or in providing evidence for new hypotheses. We will now explore some of those episodes, episodes during which mere *thinking*, in the form of thought experiments, provided good reasons for accepting certain *empirical* claim, at least provisionally.

Historically, the importance of thought experiments in the development of science can hardly be overstated. Galileo used them extensively, both to criticize the Aristotelianism that was still prevailing during his time and to establish the law of inertia, which would become a fundamental law of classical mechanics. Newton, the founder of classical mechanics, used his famous two-globes experiment to argue for the claim that space and time are absolute (i.e., entities in their own right) and not mere relations between material objects. James Clerk Maxwell employed his infamous demon to show that the second law of thermodynamics is not a deterministic (i.e., exceptionless) law, but that it is statistical in nature. In biology, Charles Darwin arguably used the results of something like a thought experiment when he applied Malthus's analysis of how resource limitations would check exponential population growth as the basis for his theory of natural selection – some organisms would be better at securing the diminishing resources and thereby survive to pass along this advantage. Albert Einstein imagined running fast enough to catch up with a beam of light in order to expose an internal inconsistency in classical electrodynamics. And there are more. In the next few sections, we will first look at the structure of some classical thought experiments, provide a classification, and discuss the evidential value and limitations of TEs.

8.2.2 Galileo's Combined Weights and Ideal Spheres

To understand Galileo's thought experiments, we need to consider the Aristotelian background which framed the prevailing theories of motion up to the sixteenth century. According to Aristotle, there are two kinds of motion[1] – natural and forced. A body moves naturally if it moves toward its natural place, which is determined by the body's composition out of the four elements: earth, water, air, and fire. Bodies with a high amount of earth tend to fall toward the center of the earth as their natural place, while bodies with a high amount of fire tend to recede from the earth – you get the idea. Forced motion, on the other hand, is any motion that is not natural. For example, an "earthy" object might move horizontally, instead of falling toward the center of the earth, if I carry it. As soon as I let go, its natural motion will take over again. Another way in which Aristotle drew this distinction was in terms of the cause of the motion – the mover, as he called that cause. His first principle was that there is no motion without a mover. In the case of natural motion, the mover is simply the nature of the object, while in the case of forced motion, the mover is some agent that acts on the object (such as me carrying a book across the room).

As you might already have noticed, certain forms of motion don't fit all that neatly into this mold of being either caused by an external mover or by the object's nature. The most problematic case was, as Aristotle himself noticed, the motion of a projectile. Not only can I carry the book across the room, I can also *throw* it across the room. What happens in that case? Clearly, the motion of the book is not fully natural. If it were, it would fall to the ground as soon as it leaves my hand. But it doesn't. Somehow, the forced motion it exhibited when it was still in my hand continues to take place for a while, even after it leaves my hand. How is that possible? Aristotle found no satisfactory solution. The first one who did was Galileo, who we'll get to momentarily.

But first, Aristotle's other major claim about motion concerned the familiar phenomenon of free fall. If I have a collection of "earthy" objects, they fall at observably different speeds – a feather falls much more slowly than a bowling ball. Aristotle postulated that the speed of falling bodies is proportional to their weight. Galileo showed that to be false by employing a thought experiment. Let's have a look at those two classic thought experiments, starting with Galileo's refutation of Aristotle's idea that speed is proportional to weight (we call it TE-1), and then his explanation for those problematical trajectories of projectiles (TE-2).

In TE-1, imagine two different objects, a bowling ball and a softball (neither ball existed in Galileo's time, but we'll stick with examples that make sense to modern life). Since the former is much heavier than the latter, it should fall much faster. To make things easy, suppose the bowling ball (15 pounds) is thirty times as heavy as the softball (0.5 pounds); however, the exact values don't really matter. Suppose further that speed is linearly dependent on weight, which means that the bowling ball falls thirty times as fast as the softball. The bowling ball falls at 30 mph, while the softball falls only at 1 mph. Now imagine that we glue the two balls together. Since the softball, by its very nature, only falls at 1 mph, it will slow down the bowling ball, from 30 mph to, say, 15 mph. However, the combined object – bowling ball plus glued-on softball – weighs 15½ pounds, so the glued-together balls should fall at 31 mph. But now we have a contradiction: The bowling ball falls both at 15 mph and at 31 mph! Therefore, Aristotle's principle that speed is proportional to weight leads to a contradiction, which means that it cannot be correct. In fact, Galileo argued that all objects, regardless of their weight, fall at the same speed, at least in a vacuum. Just imagine what would happen if any forces, such as air resistance, were absent. Why would the objects fall at different speeds? Notice the counterfactual reasoning here. During Galileo's time, it was not possible yet to produce a vacuum. Thus, he had to imagine what would happen, if it were possible to have objects fall through a vacuum, where they wouldn't experience any resisting forces.

In what we call TE-2, Galileo imagined a related scenario in his attempt to find a solution for the problem of projectile motion. Consider, he suggested, a perfectly horizontal, frictionless, and infinite plane, on which a perfectly round ball is rolling along in a vacuum. When does it stop rolling? When you think about this scenario, it becomes clear that the ball is not going to stop at all. Why should it? In the imagined conditions, there are no forces acting on the ball – no friction, no resistance, no gravitational pull (we are neglecting the curvature of the earth). Thus, it seems clear that the ball will just keep on rolling forever. In more technical terminology, the state of motion of the ball will not change in the absence of forces.[2]

The relevance of forces is even more obvious if the ball just sits there, instead of rolling along. If there are no forces, does it ever start to roll? Of course not. However, Aristotle could have accommodated that case by pointing to the absence of a mover. What Galileo's thought experiment with the rolling ball does is shift the *explanandum* – that which is to be explained. For Aristotle, what needed to be explained was motion, not rest. For Galileo, what needs to be explained are changes in the *state of motion*, of which rest is one.

This thought experiment leads to the first formulation of the law of inertia: *Any objects remains in its state of motion unless acted upon by some force*. Arguably, this law is a physical instance of a more general law, which might explain how the thought experiments works. The more general law is called the law of sufficient reason: *Anything that happens, happens for a reason that is sufficient for bringing it about*. Or, maybe a bit closer to physics: *For any effect, there exists a sufficient cause*. In light of the Galileo's explanandum, we can now say the following. What Aristotle didn't understand was the correct parsing of cause and effect regarding motion. Although he accepted a form of the principle of sufficient reason ("No motion without a mover"), Aristotle thought that only motion, and not rest, must be accounted for. He concentrated on *motion itself* as the effect that needed to be explained. Galileo, in contrast, discovered that the relevant effect is *changes in motion*, and, by the principle of sufficient reason, if there are no forces that would bring about such changes, the state of motion will remain the same – and that's the law of inertia.

This suggests that Galileo's thought experiment works because it tacitly relies on the law of sufficient reason – a law that seems to be true in light of what we mean by "cause and effect." It's a law that appears to be correct from the perspective of reason alone. It stands to reason that for any effect, there is a sufficient cause. And once Galileo identified, through his thought experiment, changes in the state of motion of an object as the relevant effect, he discovered the law of inertia.

8.2.3 Newton on Space (and Time)

The relation between our notions of space, time, and motion has been at the center of theorizing in physics for a long time. Let's take up that history with the French mathematician-philosopher René Descartes (1596–1650; he was the inventor of the Cartesian coordinate system, among other things), who wanted to find a way of distinguishing mere relative motion from "true" motion, which is needed for doing mechanics – the theory of regularities of motions and their causes. Descartes thought that mere relative motion (i.e., change of place relative to other objects) is physically unimportant, because this motion depends on an arbitrary choice of objects of reference. For example, if you drive down the highway at 75 mph, you are measuring your speed relative to Earth. Were you to measure it relative to the sun, you'd actually be speeding at an enormous rate, about 67,000 mph, with the exact value depending on your direction. To avoid this problem, Descartes proposed that the true motion of an object is that relative to its *immediate surroundings*, because there is only one immediate surrounding for each object, which means that the question of which reference object to choose doesn't arise.

Isaac Newton was also very interested in the nature of space and time. However, he thought that his theory of mechanics presupposed the notion of *absolute* space (and time). Absolute space is supposed to be something distinct from physical objects, existing independently of them. Newton didn't think of it as some sort of additional substance, but rather as the immutable arena in which all interactions between substances take place.

Contrast this with the view of the German mathematician-philosopher Gottfried Wilhelm Leibniz, who thought that space is simply the sum-total of all spatial relations among physical objects. Take the objects away, and you don't have any spatial relations left, and thus no space. Clearly, on this second conception, space is still something distinct from physical objects, but it does not exist independently of them: space is simply the totality of spatial relations, which are not objects. The relations between the objects exist only if the objects exist. This difference in their views of space (and time) was a major point of contention between Leibniz and Newton and led to bitter fights.

Moreover, this difference has consequences for how we think about motion. According to Newton, we can distinguish between *absolute motion*, which is motion relative to absolute space, and *relative motion*, which is motion relative to other objects. Imagine you are sitting in your car at an intersection, waiting for the traffic light to turn green. Suddenly you notice that the car next to you has started to ever so slowly move backward – or have you begun to roll forward? If the motion is slow and smooth enough, you might not know for a moment whether you are inching forward or the other car is inching backward. For that moment, it seems that the motion is simply a change in relative positions of the cars to each other, and without further clues, you don't know which car is responsible for the change. Leibniz thought that all (inertial, i.e., nonaccelerated) motion is like that. Newton disagreed. He claimed that there is a fact of the matter as to which one of the two cars is in motion. (In the case described he is of course right – only one of you is accelerating.) Newton thought that in every change of relative position between two objects, there is a fact of the matter as to which one is moving: It's the one that moves relative to absolute space. Thus, there is absolute motion, something Leibniz denied. As a first step in his argument, Newton relied on a physical experiment that everyone with a bucket suspended from a rope can repeat to show that Descartes' account of true motion must be wrong. However, it would be even simpler to use your background knowledge and conduct this as a thought experiment, the results of which will be quite obvious.

This experiment is simple. First notice that accelerated motion, of which circular motion is an instance, results in observable effects. Thus, if we observe any effects that cannot be attributed to relative motion, they must be attributed to absolute motion. Now take a bucket full of water suspended from a rope and twist the bucket, thereby winding up the rope (or as you read this, imagine doing so). When you let go, the bucket will spin while the water at first will remain still. After a short while, the water in the bucket will start spinning as well and eventually spin in unison with the bucket. And here is the surprising observation or realization: While the water is stationary relative to you, the observer, but is in motion relative to the spinning bucket (first stage), its surface remains flat; however, once it is spinning *with* the bucket, and thus *at rest relative to it*, its surface shows an indentation. This is clearly a motion effect that needs explanation. Figure 8.1 illustrates Newton's experiment:

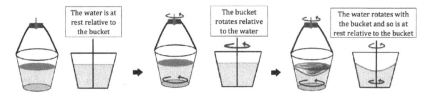

Figure 8.1 Newton's bucket experiment.

This experiment clearly shows that Descartes' identification of true motion with motion relative to an object's immediate surroundings cannot be right. The water shows a motion effect precisely when it is at rest relative to its immediate surroundings, i.e., the bucket. However, this is not enough yet to establish Newton's claim that true motion is motion relative to absolute space.

Why? Because there are plenty of other things relative to which the water is moving, including: the bar from which the bucket is suspended, the ground below the bucket, the sun, the solar system, and the entirety of the fixed stars in the universe. Thus, Newton couldn't conclude yet that the motion effects must be attributed to the water's motion relative to absolute space. There are plenty of things relative to which the water moves. Leibniz was not defeated yet.

8.2.4 Spinning Globes (TE-3)

Thus, Newton proposed a thought experiment. In it, we are to imagine a universe that is completely empty except for two qualitatively identical globes that are connected with a "rigid" cord. The cord is "rigid" in the sense that it can neither be shortened nor lengthened, and it is always taut – it never sags. The situation looks like this (Figure 8.2):

Newton's claim was this: As the two-globe system spins around its center (black vertical line) at different velocities, the cord manifests different degrees of tension. This, of course, is an effect that is in principle observable and demands an explanation. Obviously, the explanation lies in the fact that the system spins. Its varying degrees of motion around the center explain the varying degrees of tension observable in the cord. However, given that the cord is "rigid" in the sense outlined above, the distance between the globes doesn't change. Thus, they

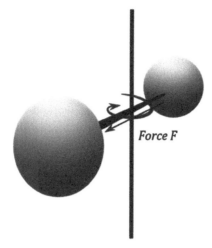

Force F

Figure 8.2 Newton's spinning globes.

do not move relative to each other, or relative to the cord. And since we made the assumption that there are absolutely no other objects in that universe, the spinning motion must be absolute motion relative to absolute space. In other words, without assuming the existence of absolute space, we could not explain the observed effects of the spinning motion.

It is important to point out that the problem with the bucket experiment (i.e., the fact that the water spins relative to the ground, the observer, etc., which removes the need to postulate absolute space to explain the effects of the water's motion) has been eliminated. By the design of the thought experiment, there are no objects relative to which the two-globe system moves. However, there are several other, and perhaps deeper, problems. First, what about that strange cord? It is supposed to be fully rigid and yet capable of exhibiting different degrees of tension. Is such a thing even possible? Second, putting aside the mysterious nature of the rope, is it really obvious that in the possible world imagined, that there would be changes in tension as a function of changes in rotational speed? Maybe once everything is removed from the universe save the two spheres, there won't be any motion effects. How would we know? We will address these sorts of questions when we discuss the evidential value and limits of thought experiments.

8.2.5 Maxwell's Demon (TE-4)

Let us finally turn to another classic thought experiment that can highlight some additional capabilities and dangers involved in using this method. This experiment commonly goes under the name of *Maxwell's Demon*. British physicist James Clerk Maxwell (1831–1879), famous for his formulation of classical electrodynamics, also worked on the nature of thermodynamic processes and their apparent irreversibility. Maybe the most well-known example of an irreversible process is the increase of entropy. In a nutshell, the law of entropy, also known as the second law of thermodynamics, states that in a closed system, entropy – a measure of disorder – increases. To see what this means, consider a container that is divided into two sections. In section A, there is a large number of molecules buzzing around at exactly the same speed. In section B, there is an equally large number of molecules that also buzz around equally fast, but at a speed greater than those in A, maybe twice as fast. If we represent the slower molecules by black dots, and the faster ones by white dots, the set-up looks like this (Figure 8.3):

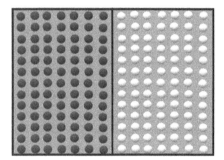

Figure 8.3 Low entropy.

Clearly, there is a lot of order in this set-up. All the black dots are on one side, while the white dots are on the other. The entire container is nicely ordered into two sections of similar objects: black and white dots.

Now imagine that we remove the wall between sections A and B. What will happen? Clearly, we will lose the order, and eventually end up with a random distribution of molecules at different speeds. If we could see molecules, the contents of the container would look like this (Figure 8.4):

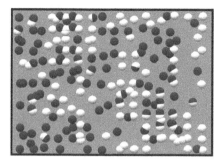

Figure 8.4 *High entropy.*

This illustrates, roughly, the idea of increasing entropy. In a closed system, order is inevitably lost with the passage of time. Closed systems evolve from states of low entropy (high orderliness) to states of high entropy (low orderliness). In the early nineteenth century, many physicists thought that this law of entropy holds universally and necessarily. This is where Maxwell's Demon comes in. With his thought experiment, Maxwell tried to show that the law of entropy does not hold universally and necessarily, but only with high probability. He tried to show that it was only a *statistical* law, much like there being nothing physically necessary that precludes every coin flip being heads for the next hundred years.

To see how this works, notice first that the two pictures above can also be used to illustrate a particular instance of entropy, namely heat flow. In a closed system, heat always flows from warm to cold. If you put a hot object in contact with a cold one and leave them touching each other, they will eventually reach thermal equilibrium – they will have the same temperature. What you definitely won't see is that the hot object gets hotter and the cold one gets colder. And this seems necessary.

But is it? The top picture shows a cold object (black dots) in contact with a hot object (white dots). According to the kinetic theory of heat, the *heat* of a gas simply is identical with the *mean molecular kinetic energy* of the gas molecules. That's why we can represent the cold object with black dots, which are slowly moving molecules, and the hot object with the white dots, which are molecules moving much faster. However, at this point it is crucial to emphasize that heat is the *mean* molecular kinetic energy. It follows that in a hot object, most molecules will move fast, but there will also be some slow ones, and in a cold object, most molecules will move slowly, but there will also be some fast ones. The law of entropy, in this context, simply says that the white dots from the hot object will flow into the cold object, and the black dots from the cold object will flow into the hot object. Thus, after some amount of time, the resulting situation will be exactly the random mixture of black and white dots illustrated in the picture above. Entropy wins. Order is lost. And this seems necessary.

Now we can look at Maxwell's thought experiment. In a nutshell, he told us to imagine a trap door in the wall between the two sections of the random mixture. Next to this is a demon that only opens the door when a black dot arrives from the right or white dot from the left, the end result will be perfect order. In other words, the demon can reverse entropy, and move from the bottom picture to the top one. It looks like this (Figure 8.5), replacing the demon with a cat (which is arguably not much of a change):

Figure 8.5 Maxwell's "demon".

Clearly, we have moved from disorder (high entropy) to order (low entropy). Although we have never observed such a violation, the law of entropy doesn't hold with necessity. It is possible to reverse the trend (by chance, if not feline demons, as we shall see), although it is extremely improbable in our universe. But this doesn't change the fact that the law of entropy is only a statistical law, and not one that holds universally and necessarily. This thought experiment is used to *correct an inductive inference* we made on the basis of our observations. We have only observed increases in entropy, and never decreases. Thus, we concluded that in *all* closed systems, entropy increases over time. What Maxwell's thought experiment seems to show is that this inductive conclusion is false. There are conceivable cases in which entropy decreases over time.

There may be a glitch though. What about the demon? The problem is not that demons don't exist. We can grant Maxwell this element of the thought experiment because we can agree that demons are possible and thus there is a possible world in which they do just those things pictured above. However, the question is whether the demon's actions change the system from a closed system to an open one. This is important because we know that the law of entropy doesn't hold for open system, which are systems that receive energy from the outside. After all, we have refrigerators, in which already cold objects get even colder. But refrigerators are not closed systems – they need energy to accomplish the feat of making cold objects colder (and the air just behind the appliance hotter). So we have to ask about the introduction of the demon into a system in which two chambers are separated by a wall with a trap door: Isn't there a sense in which the demon, by opening and closing the door in appropriate ways, brings energy to the table, and it is this fact which explains that entropy decreases?

8.3 The Structure and Evidential Value of Thought Experiments

8.3.1 Kinds of TEs

There are vigorous debates in the literature about whether TEs are really anything more than embellished forms of otherwise ordinary reasoning (TEs as *imaginative reasoning*), or whether they can provide some evidence for empirical claims (TEs as *mental evidence*). There are also debates about whether or not TEs can only be used for destructive purposes by, for example, exposing some fatal flaw in a theory, or whether they can help in constructing and further articulating theories. Without trying to resolve those debates, it is useful to try to categorize TEs in accordance with the possible views on the structure and role of TEs, remaining cognizant of the fact that the categorization is largely conventional. Accordingly, we distinguish between TEs as forms of imaginative reasoning and TEs as mental evidence along one dimension, and between destructive and constructive TEs along the other dimension. Dimensions are preferred to discrete categories, as they don't suggest fixed and exact boundaries.

8.3.2 TEs as Imaginative Reasoning

Some have argued that certain TEs are simply instances of ordinary reasoning guided by visualizable scenarios. These forms of reasoning are imaginative in that they involve mental imagery that perhaps facilitates the reasoning. They can be used for both destructive and constructive purposes. A good example of destructive imaginative reasoning is Galileo's TE-1 about freely falling objects. According to Aristotle, the speed of a falling object is proportional to its weight: $s = f(w)$. Suppose one object is twice as heavy as another object: $w_1 = 2 \cdot w_2$. Then, $s_1 = 2 \cdot s_2$. If the two objects are glued together, one will retard the other, while the other will speed up the one. Therefore, the common speed should be 1.5 units. However, the resulting object is also heavier than the two individual objects ($w_1 + w_2$), so that their common speed should be 3 units. Thus, Aristotle's theory entails a contradiction and cannot be correct. However, that we glue balls and feathers, or any other objects, together in the TE only helps us keep track of, or illustrate, the math. The argument itself is largely mathematical, together with the extra assumption that the speed is fixed by the weight, so that we get both decrease and increase of speed when combining two objects in free fall.

Maxwell's Demon (TE-4) may be described as a case of imaginative reasoning put to constructive purposes. He developed it before the background of a fairly well-articulated theory (classical thermodynamics) and used it to support a claim that is perhaps not obvious from the austere mathematical formulation of the theory. The claim was that the law of entropy is statistical rather than deterministic (given the latter, the TE can also be seen as partly destructive). The identification of the temperature of a gas with the mean molecular kinetic energy of its constituents entails, together with the absence of any deterministic laws dictating the direction of movement for individual particles as a function of their speed, the physical possibility that all particles above a certain threshold of speed end up in one of the connected chambers, while the particles with speeds below that threshold end up in the other one. Introducing a demon who arranges for that outcome simply makes the possibility more vivid, but it is strictly speaking not required for appreciating the physical possibility of decreasing entropy. This observation also solves the alleged problem we hinted at earlier, viz., that the demon seems to introduce a source of external energy, rendering the system of the connected chambers in an important sense an open system. But the demon can be dispensed with and we get the outcome purely by chance. The physical possibility of a reversal of entropy is guaranteed by the random distribution of speeds among the particles and the absence of any law regulating the movement of the individual particles that is sensitive to those speeds. The demon helps to see this, but the demon is not essential to Maxwell's argument.

8.3.3 TEs as Mental Evidence

Let's start with Galileo's TE concerning the law of inertia (TE-2). In an important passage of his *Dialogue Concerning the Two Chief World Systems*, the character Salviati asks his interlocutor Simplicio what would happen if a perfectly round ball rolled down a downward sloping plane – would it ever stop without a force being applied? Alternatively, would it ever roll up a sloping plane without a force being applied? Simplicio responds with a "no" to both questions. Then, when asked what would happen to the ball on a horizontal plane, he answers that there would be "neither a natural tendency toward motion nor a resistance to being moved" and thus it would remain at rest. Finally, when asked what would happen to a moving ball on such a plane, Simplicio answers, in effect, that since there is no cause for stopping it nor for accelerating it, the ball would remain in the same state of motion *perpetually*.

   This Galilean TE is clearly different from the TE he used against Aristotle's theory of free fall. It seems to be more than simply math with imagery. Rather, this is an instance of counterfactual reasoning. The relevant counterfactual is: "If there were a perfectly round ball rolling on a perfectly smooth and horizontal plane, it would never stop rolling." It is true just in case that in any scenario in which a perfectly round ball rolls on a perfectly smooth and horizontal plane, it doesn't stop rolling. Simplicio is led to conclude that this counterfactual is true by two pieces of information: (i) if a rolling ball on a downward slope is to be stopped, a force needs to be applied and (ii) if a ball sitting on an upward slope is to be set into motion, a force needs to be applied. In the antecedent of the counterfactual, we are to imagine that both upward and downward directed forces are absent. Then, ignoring (abstracting away from) all other forces, such as friction and gravity, we conclude, with Simplicio, that if the ball is at rest, it will remain at rest, and if it is moving, it will remain moving.

   Perhaps this TE is very similar to Newton's two-globe experiment (TE-3). There, we abstracted away from any reference bodies for relative motion, which results in the claim that the two globes must be moving absolutely. In the present case, we may think of situations in which fewer and fewer forces are present as approximating a situation in which no forces are present. If successive reductions in the number of forces acting on a moving body leads to successively longer periods of motion, then taking the number of forces to zero must mean that the motion goes on forever. In both cases, then, we extrapolate from ordinary experience to what would happen in extraordinary circumstances – we extrapolate from the real world to a counterfactual world. Under which conditions this process is legitimate is the topic of the next section.

Before that, let's quickly take stock. We distinguished between constructive (C) and destructive (D) TEs, and between TEs as imaginative reasoning (IR) and as mental evidence (ME). The following graph (Figure 8.6) sums up our discussion of the classic examples relative to those dimensions.

8.4 Learning from TEs

Can scientists use the results of thought experiments to gain an understanding of the real world? How can merely thinking about the world reveal anything about reality? Looking at the space of possibilities above, it seems that TEs in the left half are the least problematic. In particular, TE-1, Galileo's attack on Aristotle's theory of free fall, is a case of ordinary mathematical reasoning supported by mental imagery. We can learn from it that Aristotle's theory must be wrong, because it entails a contradiction. Maxwell's Demon (TE-4) seems equally unproblematic. As was pointed out above, the statistical character of thermodynamics, and of the law of

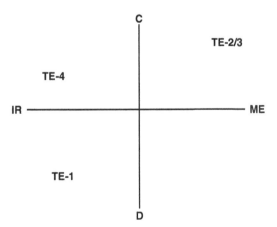

Figure 8.6 Classification of thought experiments.

entropy in particular, follows from the theory itself, and the demon simply has the purpose of making this fact more vivid. The tricky cases are those located in the upper right quadrant, Newton's two-globes experiment (TE-3) and Galileo's smooth balls on smooth planes (TE-2).

8.4.1 Distant Worlds

When we imagine possible worlds, it is clear that some of them are much more similar to our actual world than others. For example, a world in which Joe does not spend all his money on clothing, and thus has enough left over for food, is fairly similar to the actual world, even though here, Joe does spend all his money on clothing and consequently goes hungry. At any rate, the "frugal-Joe-world" is certainly more similar to our world than a world in which an angel gives Joe increasingly greater amounts of money, the more he spends on clothing. In the literature on possible worlds, overall similarities between worlds are expressed in terms of *distances*. A world is nearby if it shows a great degree of similarity with the actual world, and it is distant if it is very dissimilar from the actual world. Thus, the frugal-Joe-world is much closer to our world than the angel-world.

TEs involve considerations of other possible worlds. Sometimes, those possible worlds can be made actual. Galileo's world in which we tie two falling objects together can easily be made actual. However, some worlds are distant enough that they can't be made actual. Galileo's world of perfectly smooth balls rolling in the absence of any forces on perfectly horizontal and frictionless planes of infinite length is not a world that we can make actual. There simply are no perfectly smooth balls rolling on perfectly horizontal planes that are infinitely long to be found anywhere in the actual world. The same goes for Newton's world that is absolutely empty but for two globes, connected by a rigid but tension-changing cord, that spin around their common center of gravity. Just try to remove the entirety of the fixed stars!

The deep problem is this: Should we infer anything about what would happen in such distant worlds on the basis of what happens in related situations in the actual world? In other words, should we really infer that in Galileo's world of perfect balls and planes, those balls would roll forever? It seems quite risky to make extrapolations about motion in perfect worlds from motions in the actual and imperfect world. It is perhaps even riskier to infer anything about Newton's two-globe world, which is about as distant as it gets. Does the effect he postulates really occur there? A couple of considerations would suggest that we can't really know.

That there is an observable effect has to be taken with a grain of salt. In one sense, the varying degrees of tension on the cord count as observable in principle – for if we introduced an observer into the universe, she would observe the tension (by feeling the cord, say). However, introducing such an observer would also destroy the experimental set-up, because now there would be an object relative to which the two-globe system spins (viz., the observer we just introduced) so that the assumption of absolute space is no longer necessary. Maybe what we need, then, is to introduce an observer that has no position in space and time, some sort of abstract observer (abstract objects, such as numbers, are not located in space and time).

Given, however, that observation is, for all we know, grounded in causal interactions, and causal interactions are interactions between objects located in space and time (a real observer must touch the cord to feel the tension, something an abstract observer couldn't do, because it is not located anywhere, let alone close enough to the cord to be able to touch it), this kind of abstract observation would be extremely unfamiliar and might well be impossible. If it is impossible, then it can't happen in the two-globe universe, which means that the alleged changes in tension would not be observable effects that need to be explained.

Even if we put aside the problem with the observability of the effect and simply rely on a sort of *modal intuition*, there is another problem. If one were to say that it is clearly possible that the tension changes occur in a two-globe world – after all, there is no contradiction in supposing it does – the claim that absolute space exists in the actual world would not be supported. At best, it would show that the notion of absolute space, which manifests itself in mechanical effects of motion that is not motion in relation to other objects, is not contradictory, and thus that absolute space is possible. But this is of course a far cry from what Newton wants, namely, to show that absolute space is actual, or real. TE-3 seems doomed.

Does the same follow for TE-2? Given that Galileo's world of perfect balls is also extremely distant (but perhaps less so) from ours, do we have a good reason for supposing that those balls would roll on forever? Perhaps the "motive force" imparted to the balls exhausts itself over time, and so the balls would eventually come to a stop. (This theoretical possibility has actually been proposed as a solution to the problem of projectile motion by medieval defenders of the so-called *impetus theory*.) In our world, other forces bring the rolling ball to a stop long before the motive force is exhausted. But in Galileo's world, the effect of exhaustion might take over fully. Is there any way to save the law of inertia introduced by the TE, or does it experience the same fate as Newton's TE-3?

8.4.2 TEs as Tools for Discovery

Fortunately for Galileo, there *is* one important difference between TE-3 (space) and TE-2 (inertia). The law of inertia was eventually incorporated into classical mechanics, a theory that was more powerful – both in its explanations and in its predictions – than anything that had come before. Thus, even the results of thought experiments can, under the right circumstances, be subsequently tested using a physical experiment when they become embedded in theories that make empirical predictions.

Maybe the lesson to be learned here is this. TEs as mental evidence are useful for suggesting certain theories or hypotheses about the empirical world, but they ought not be used in an attempt to justify those theories or hypotheses. In other words, TEs from the upper right quadrant are *tools of discovery*. Both the two-globe-world and the perfect-balls-world suggest to us certain theories about reality. However, they do not justify them. Instead, their justification is arrived at in the usual way, viz., by showing that by incorporating the results from a TE we can develop a powerful theory that is empirically confirmed through ordinary experiments and other tests. In the terminology of Chapter 3, TEs from the upper right quadrant belong to the context of discovery. When we move to the context of justification, we see that the claim (i.e., the law of inertia) derived from TE-2 can be justified by being embedded in a testable theory. It is unclear whether any claims concerning absolute space can be equally justified in this manner.

8.5 The Ubiquity of Thought Experiments

For many students of science, thought experiments likely seem esoteric. Few science textbooks mention this approach to exploring the natural world. But there is an important sense in which thought experiments are the most commonly employed approach in all of science. Let's say that you've observed some intriguing phenomenon: a strain of hamsters that appear to live without ever imbibing water, a pendulum that seems to have an increasing period through the first several swings, or a tribe of people who evidently communicate without words or gestures. How would you proceed with regard to seeking explanations? You might try drying the hamster's food to see if they derive water metabolically, or isolating the pendulum from the potential of magnetic fields to ascertain if some external force was at work, or blindfolding the tribal members to determine if they use subtle, visual cues.

The point is that no matter how you design a physical experiment, you begin by imagining what might happen if your hypothetical treatment explained the phenomenon. That is to say, you begin with a thought experiment! Indeed, this seems to be exactly what Newton did with his bucket-of-water experiment that he didn't even need to actually perform, but which we can quite easily realize. And in some cases, it might transpire that when you thought through the planned experiment, you realized that something was missing, that some other factor could confound the results, or that another control would be more effective. Perhaps you want to put the hamster in a low humidity cage to exclude the possible absorption of water through its skin or maybe you'll want to plug the noses of the tribal members to exclude the possibility of pheromonal cues. Scientists involve their imaginations and conduct thought experiments on a frequent basis. They are, in a sense, very effective thought experimenters.

Thought experiments are a kind of fiction, so following this line of analysis, should we consider short stories, novels, plays, and films as thought experiments? In a sense, the writer is proposing and exploring a possible world, varying certain features (treatments), and comparing the results to what would have happened without those variations (this is like imagining a temporal control). It does seem that we can learn about psychology and social behavior through works such as *Sophie's Choice*, *Othello*, *1984*, and *Brave New World*. Even nonfictional stories can become thought experiments at the individual level, as we read the tale of a stranded group of travelers and imagine whether, when, and how we'd engage in cannibalism (e.g, the book, *Alive: The Story of the Andes Survivors* and the subsequent film). We do not wish to belabor this extension of thought experiments to the page, stage, and screen, but we suggest that it would be a meaningful exercise to explore the conditions under which a work of fiction would plausibly constitute a thought experiment in the sense of scientific inquiry.

8.6 Are Computer Simulations Thought Experiments?

Computer simulations are also experiments, sitting somewhere between a thought experiment and a physical experiment, bearing additional similarities with scientific models (see next chapter). To see their kinship with TEs, let's look at some parallels between these two scientific tools by way of closing the current chapter.

A central feature of thought experiments involves our capacity to imagine situations that can't be instantiated in the world but which might reveal important aspects of reality. While we can't conceptualize what isn't logically possible (e.g., a round square), we can conceive of possible worlds, as noted earlier. But this venture is not just a method for experiments taking place in our heads, it increasingly provides a potent basis for a kind of experiment that is performed in the world – or at least the artificial world of computers.

Computer simulations allow us to create models of the world and then systematically alter features to infer what factors influence the behavior of complex systems. The alternative realities that we create via computers can provide a means of exploring aspects of the world that we can't otherwise access because of spatiotemporal constraints (i.e., very long periods of time as with evolutionary simulations or extremely large scales as with climate change), economic limitations (e.g., the effects of building a dam to ascertain changes to a river's hydrology) or ethical concerns (e.g., testing how a novel virus would move through a population and whether children and seniors would experience disproportionate mortality with and without vaccination). And in some fields, such as high-energy physics, simulations are an essential part of all experiments.

One of the interesting questions in this context pertains to the metaphysics of simulations: What is their nature, how are they related to mere possible worlds, and how are they related to the physical world? Perhaps it is plausible to assume that the world in a simulation is simply a very nearby possible world, one in which the stand-ins for physical processes, for example, take place on (much) smaller temporal and spatial scales. Or they are nearby worlds devoid of actual agents that can experience harm, but still be subject to simulated diseases. Much like thought experiments, simulations are certainly cheaper than many physical experiments, although not quite as cheap as pure TEs. And, like scientific models, they can mediate between high-level theory and the physical world by producing data that are similar to those we would receive from a physical experiment. Thus, computer simulations are, to a first approximation, dynamic models that realize nearby possible worlds and help in this way to mediate between high-level theories and physical reality to provide a context for experimentation.

8.7 Conclusion

TEs are certainly not part of standard laboratory manuals in university science courses, which might lead one to infer that these strange ventures are not vital to the practice of science. But as we've shown, TEs were not only essential for a number of important scientific break-throughs in the past, they are also fundamental insofar as the scientist imagines what might happen with a particular intervention in the natural world during the design of a physical experiment. Moreover, the line is blurred between a TE and a simulation, the latter being increasingly important to scientific research. It seems that students of science would be well advised to reflect on the power of our "laboratories in the mind" (or in a computer), appreciating how they can supplement or at least guide actual, physical experimentation.

Notes

1   More precisely, there are two kinds of locomotion, which Aristotle thought of as a species of change, i.e., change in place. However, we simply use "motion" instead of "locomotion".

2   To be sure, Galileo didn't talk about forces; in fact, he was not interested in causal accounts at all, and forces are kinds of causes. But our anachronistic way of describing the TE doesn't change its logic.

Annotated Bibliography

James Robert Brown, 1991, *The Laboratory of the Mind. Thought Experiments in the Natural Sciences*. New York and London: Routledge.
    Brown defends the view that some TEs allow us to directly grasp certain abstract entities important for science, namely, laws of nature.

Roy Sorensen, 1992, *Thought Experiments*. New York and Oxford: Oxford University Press.

The author proposes that thought experiments are a part of unexecuted experiments, which often probe the modal implications of various theories and thereby have the capacity to test them.

John D. Norton, "Why Thought Experiments Do Not Transcend Empiricism," in Christopher Hitchcock (ed.), *Contemporary Debates in the Philosophy of Science*, Oxford: Blackwell, 44–66. Preprint available at http://philsci-archive.pitt.edu/960

Norton argues that thought experiments are just arguments supported by visual imagery which is strictly speaking not required for establishing the relevant results.

9

MODELS: USEFUL LIES AND INFORMATIVE FICTIONS

In this chapter, we will discuss the important concept of a scientific model. Constructing and using models – in short, the activity of modelling – has increased in importance over the last decades; at the very least, using the terms "models" and "modelling" has become more widespread over the years. Thus, it is important for the philosopher of science (and the scientist, of course) to understand what's involved in modelling, and to determine how models and modelling are related to the goals of forming true and well-justified beliefs about the world. The problems posed by models for our attempts to understand the world are particularly challenging. As we will see, all models involve simplifications, and some simplifications are idealizations. In general, simplifying has the effect that we do not represent all features of a system that we try to understand, while by idealizing, we often introduce assumptions into a model that we know are false. In what sense, then, can we be said to learn anything about the world – to acquire true beliefs about it – on the basis of simplifying and false modelling assumptions? Put somewhat provocatively, if models are fictions, how can they give us the truth about the world? (We asked the same question about thought experiments.)

In order to answer those questions, we will have to first achieve a better understanding of the nature of models and how idealization, simplification, approximation, and similar techniques are used in their construction. To reach this goal, we are going to try to find commonalities among familiar examples of models and then see whether those commonalities extend to

This is Philosophy of Science: An Introduction, First Edition. Franz-Peter Griesmaier and Jeffrey A. Lockwood.
© 2022 John Wiley & Sons, Inc. Published 2022 by John Wiley & Sons, Inc.

most, if not all, models that are employed in the sciences. In this way, we hope to arrive at something like an account of the nature of models and their use.

After isolating the commonalities among scientific models, we discuss various techniques used in model construction. Those techniques include different forms of simplification, such as abstraction, approximation, and idealization. We concentrate on those techniques because they raise the most interesting and important questions about how we can learn something about real-life systems (from now on called *target systems*) through modelling. Each technique is connected with different sets of issues. For example, scale models give rise to questions about which features of a target system can and cannot be profitably scaled up (or down). Idealizations, on the other hand, involve us in discussions about modal distances between the world of the target system and the idealized world of the model and how these distances affect what we can learn from the model. We already encountered a similar issue in the last chapter on thought experiments. This alone suggests a strong kinship between models and thought experiments.

Modelling is often seen as a somewhat esoteric, scientific specialty, but in reality, all of us have been familiar with modelling since early childhood. Dolls, toy cars, train sets, even action figures – all of those are (scale) models of other systems. Of course, their primary purpose was to entertain us and provide pleasurable afternoon hours. But even playing during childhood with one's toys can, and often does, result in learning new things: Building a sandcastle, e.g., allowed us to learn rudimentary facts about statics, soil stability, and design, to name just a few. On the other hand, scientific modelling is serious work but can be enjoyable at the same time. However, can we learn something about scientific models from our experience with toys?

The answer is a qualified yes, at least when we characterize models in a very abstract way. From this perspective, models are simply *stand-ins* for target systems, which are the systems about whose features we are trying to learn something. For example, playing with model train sets can allow us to learn something about how schedules have to be designed in order to avoid collisions. A particular train set might in this way be a stand-in for a section of tracks in a country's public transportation system. Since it can be difficult to schedule departures and arrivals of many different trains that travel on common and often intersecting tracks by just looking at time tables, exploring a particular scheduling system using a model may help avoid dangerous situations and even catastrophic collisions. Today, this is largely done by computer simulations, but it could be done with scale models, such as a model train set, as well. Clearly, playing with toys is also closely related to simulations, just that in this case, the simulations are run on physical stand-ins for the target system.

9.1 The Nature of Models

We start with an example consisting of two maps, a street map of the inner city of Vienna, Austria, and a map of the Vienna subway system. As you'll see, there are commonalities between them, but also important differences, which are mainly grounded in their intended uses. Here is the first map (Figure 9.1).

This is a map of the first district in Vienna, Austria. It's where many tourist attractions are located. Toward the upper left corner, you read "Schottentor," which indicates the subway station next to Vienna's main university. You can get a lot of additional information from this model of the system, i.e., Vienna. It retains the spatial relations among the streets, and also the spatial distances between various points of interest. Thus, not only can you find out which combination of streets will get you on foot from the subway

Figure 9.1 Vienna street map, selection[1].

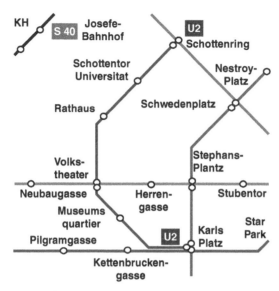

Figure 9.2 Vienna subway map, selection[2].

stop "Schottentor" to the Stephansplatz (toward the center; this is where the famous St. Stephen's Cathedral is located), but because the map encodes distance information, you can also find the shortest trip.

Now let's look at the subway map (Figure 9.2). As you can see, there is still some of the same information as there is on the street map, but lots of details are left out.

The subway stop "Schottentor" is still indicated on this map but without any real distance information from there to the Stephansplatz. The subway map shows only the sequence of the stops and indicates at which stops you can switch trains. Information about distances is not included in the subway map, and neither is information about the spatial relations among different streets, because it is not important to represent real distances in the subway map. After all, once you are on the subway, you can't decide to get off nearer to your ultimate destination than the nearest stop. That is of course not the case when walking to your destination, and a map giving you information about actual distances is much more useful. Thus, it is sufficient for a subway map's usability that it encodes the sequences of stops, while the street map should also encode actual distances. This is a difference between these maps that derives from their different intended uses.

Thus, the intended use of a model plays a big role in determining its properties. For example, models of real-life systems leave out a lot of details – if we included literally *all* features of the system, we would end up with a replica of the system and not with a model. However, which details can be left out is determined by the purpose of the model.

So what are the commonalities between our two maps? Both are obviously stand-ins for the real thing, one for the subway system, the other for the physical layout of part of the city – streets, intersections, etc. Equally obviously, there is a clear sense in which they are maps of the same entity: Vienna. The two maps simply represent different aspects of Vienna, but ultimately, *by* representing aspects of Vienna, both *partially* represent Vienna. Perhaps, then, we should think of models as representations of relevant target systems. This idea can be encapsulated in the following proposal:

(1) A scientific model is a representation of a real system in the world (its target system).

Clearly, this is too broad and doesn't exclude sufficiently many things from being models. The word "grandmother" represents your grandmother, but it surely is not a model of your grandmother in the same sense in which the double-helix is a model of DNA. A word represents in the sense that it specifies, e.g., what a sentence is about, as in "My grandmother baked these cookies." But the word models nothing. The term "representation" seems to be too broad. While all models are representations, not all representations are models. What could we add as a further condition?

Thinking about the example of the maps, they have features that are lacking in the word "grandmother." The maps are somehow *similar* to their targets. In particular, there are certain kinds of *structural similarities* between map and target. The street map shows the structural layout of streets in that part of Vienna. What it doesn't show are other features of the structure that is Vienna – the locations of sewers, e.g., which are important to the city maintenance department, or the locations of restaurants which might be vital to tourists. Such partial structural similarities are called *partial isomorphisms*. Being partially isomorphic to a target system might be what all models have in common.

9.1.1 Models as Partial Isomorphisms

Many models in the sciences are at least conjectured to be partial isomorphisms. One of them is the Bohr model of hydrogen (Figure 9.3):

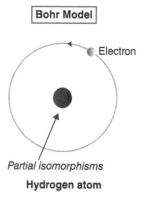

Figure 9.3 Bohr's hydrogen model.

Of course, nobody claims that this model is similar to the hydrogen atom beyond structural properties. The electron is not a little round ball that orbits another little ball, called the nucleus. What matters is the existence of a distance between the electron and the nucleus, and the further claim, seen in this more elaborate and general model, that there are only certain allowable orbits or shells (Figure 9.4):

It seems, then, that typical models represent structural features of their target systems. Here is a second, more precise, characterization of models:

(2) *A model is a structure that is partially isomorphic to its target system.*

This proposed characterization contains at least two terms that need further unpacking: the term "structure" and the term "partially isomorphic." In the example of the Bohr model, what we mean by "structure" is pretty self-explanatory. They both show visual structures that are supposed

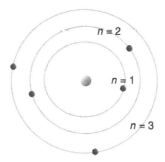

Figure 9.4 Electron shells.

to represent the structure of the atom. The second version with electron shells contains additional information about electron orbits in general, but this information is also structural information about the allowable spatial relations between the nucleus and electron orbits. At any rate, the only properties of the target systems represented by either model are structural properties. In the examples, these are structural properties that can be easily visualized.

There are other ways in which structural properties can be represented. For example, we can represent the spatial layout of some system in terms of the Cartesian coordinates of the elements of the system. Instead of drawing two points of the system next to each other with a definite distance between them, we could simply provide the Cartesian coordinates of the points: A = <3, 3, 5> and B = <3, 3, 6>. This would mean that point B is located one unit of distance away from A along the coordinate z. Although this sort of representation is a bit more abstract, we can still capture it by (2). This becomes clear after we spell out the notion of an isomorphism.

An isomorphism is a mapping from the elements of one structure onto the elements of another structure such that the relations among the elements are preserved.[3]

The Bohr model of the hydrogen atom is (conjectured to be) isomorphic to the hydrogen atom. It is important to point out that it is, of course, only partially isomorphic. Some information – for example, information about the spatial extensions of the electron and the nucleus – is not being represented. This raises the question about the rules that determine which information should be represented in a model. The short answer is: There are no strict rules. The structural relations we include in our model are determined by our interests, which are often tied up with the intended use of the model. We have already seen this in our example of the two maps of Vienna.

The fact that interests, through intended use, determine which aspects of a real system we want to model can be captured in the following modification of (2):

(3) *A model is a structure that is partially isomorphic to its target system by being fully isomorphic to those substructures in the target system that answer to the intended function of the model.*

Since (3) is somewhat cumbersome, we will simply say:

(4) *A model is a functional structure that is partially isomorphic to its target system.*

In light of (4), the subway map is a model of those structures in Vienna that answer to the map's intended function: to help navigate the subway system. The city map is a model of those structures in Vienna that answer to this map's intended function: to help navigate the layout of the city's streets.

Thus, (4) raises the question about what sort of functions a scientific model is typically intended to have, because as it stands, (4) allows things like board games and theater plays to count as models. Consider a game of Monopoly. It involves a dynamic structure that is isomorphic to that of selling and buying real estate in a largely unregulated market. However, we are reluctant to call the game a model of real estate markets. Playing a game is different from the activity of modelling. In modelling, we often want to learn something about a system that we could otherwise not learn. For example, in applying a predator–prey model, we might seek an explanation of the otherwise mysterious decline in the number of certain animals. Moreover, even if we don't know at present that there are predators in the system, applying the model and noticing that it predicts the observed rates of disappearance quite accurately might prompt us to look harder for a predator, and the rates of disappearance might even provide further hints as to what sort of predator we are looking for. These considerations suggest that one of the typical intended functions of a model is to help us understand features of its target system.

9.1.2 Mirror Models vs. Conjecture Models

Unfortunately, (4) has at least one serious shortcoming. Consider René Descartes's "screw model" of magnetism. Briefly, in order to explain the otherwise mysterious magnetic attraction between a loadstone and small pieces of iron, Descartes postulated that this "attraction" is the result of screw-shaped particles in the ether (a hypothetical fine substance pervading the entire universe). According to this model, there are certain pores in loadstone, through which the screw-shaped particles fit, and once they latch onto a piece of iron, they start to turn so that iron and loadstone approach each other (think about securing one object to another by putting a screw through both and then tightening the whole thing with a nut). Thus, Descartes didn't need to postulate any mysterious force that acts at a distance, as he was able to postulate a mechanism that accounts for magnetic attraction.

This poses a problem for the analysis of models provided by (4), which claims that a model is a functional structure that is partially isomorphic to its target system. Now the problem is not only that Descartes's "screw model" is not even partially isomorphic to its target structure (interactions between loadstones and iron), but *that we don't know what this structure is, because we cannot observe it!* In other words, (4) presupposes that the structures in the target system that we model are observable and can be used to guide model construction. This is true for many models, such as maps. However, when we produce models that involve unobservable elements, we can't be guided by observed structures in the construction of our models. We don't see those structures!

We already have seen examples of such cases. Bohr's model of the atom is one such case, and the predator-prey model, in a situation in which we haven't observed the predator, is another case of this sort. Let us call such models, in which the model structure is not directly derived from observed structures, *conjecture models*, and let us call the other class of models, where we can be guided by observed structures, *mirror models*. We can now replace (4) by the following:

(5) *A mirror model is a functional structure that is partially isomorphic to its target system.*

(6) *A conjecture model is a functional structure that is conjectured to be partially isomorphic to its target system.*

Thus, (5) should be pretty obvious, as it only introduces the new modifier "mirror." On the other hand, (6) requires some further comments. First, what is the typical function of a conjecture model? The function of Descartes's model was to explain a certain regularity (magnetic attraction). The function of Bohr's model was also to explain certain regularities (the so-called Balmer series in the emission spectrum of hydrogen; don't worry about the details). In general, we might say that conjecture models are the results of *inference to the best explanation* of observed regularities.

Second, conjecture models are not always taken to be literally true, or even approximately. While it's unclear whether Descartes thought of his model as approximating the truth, there have been many models proposed by scientists that they warned should not be taken as making truth claims. For example, Newton claimed about his model (called Classical Mechanics), which introduced the law of universal gravitation, that it is not intended to advance a "physical hypothesis" about the world. Similarly, James Clerk Maxwell thought of his models as mere *analogies*, not to be confused with a truthful descriptions of structures underlying the observed phenomena. If some conjecture models are not even supposed to mirror actual structures underlying the phenomena observed, we need to ask how we construct them and in virtue of what we judge them as successful. Before we turn to this difficult task, let's focus once more on mirror models and try to understand how the techniques of simplification, approximation, and idealization are involved in their construction, and what consequences their involvement has for the epistemic status of those mirror models. If mirror models involve fictions, how can we learn from them?

9.2 Modelling Techniques

Typical models incorporate a reduction of the information – or perhaps more accurately, an increase in the signal:noise ratio – such that the lost "information" allows a gain in our understanding of the target system. Maps are a good example. A typical city map does not include information about the location of sewers. We suspect that some models reduce information, not just for reasons of practicality, but also due to the desire to have a model that is generally applicable and thus in need of being purged of many elements that only figure in the particular system at hand. In this way, information reduction could be seen as analogous to the choice of simple curves as an answer to the problem of overfitting (a connect-the-dots approach is very precise but doesn't result in a generalizable curve).

Whatever the modelling goal, we now turn to the question of how this reduction is achieved. The principal method is *simplification*, which itself can be divided into *approximation* and *idealization*. In the literature, we find a further distinction between two forms of idealization, viz., Aristotelian and Galilean. Aristotelian idealization consists mostly in *abstracting* away from the properties of a system that seem irrelevant to the behavior of interest,

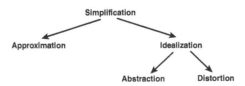

Figure 9.5 Modelling techniques.

while Galilean idealization introduces deliberate *distortions*. Simplification in model construction thus looks roughly like this (Figure 9.5):

9.2.1 Approximation

The notion of approximation is most familiar from mathematical contexts. For example, an approximate value of π is 3.14159. In antiquity, mathematicians used an approximation technique for determining the area under a curve. They simply divided the area into ever smaller rectangles (whose areas can easily be calculated) and took the sum of their areas. In modelling, we often approximate the equations governing some aspects of a real system by a model (i.e., equation) that makes some simplifying assumption. For example, suppose you want to model the motion of a heavy body through a very viscous liquid (such as oil). As the body moves through the liquid, it drags with it some of the liquid's molecules, making the situation one in which we have an interaction between several objects: the body, the liquid, and a number of dragged-along molecules. This results in dynamical equations that are difficult, if not impossible, to solve. Thus, we use what's called *renormalization* and simply add the mass of the dragged molecules to the mass of the body. Now we have to just solve an equation describing the interaction between one object (the body with its renormalized mass) and the surrounding liquid.

9.2.2 Abstraction (Aristotelian Idealization)

Suppose you want to model the relation between the momenta before and after a collision between the cue-ball and some other ball. It turns out that (to a very good approximation), the sum of the momenta before the collision equals the sum of the momenta after the collision. You now model the billiard game as a system in which the sum of momenta is preserved across collisions. Clearly, there are a lot of aspects of the game your model abstracts away from: the patterns on the balls (striped vs. solid), the color of the felt, the lighting conditions, etc. We do this because their influence on how the balls behave during a collision is negligible or nonexistent. Of course, if we want to model the strategy of the game, the patterns on the balls become paramount.

The ideal pendulum law $\left(T = 2\pi\sqrt{l / g} \right)$ abstracts away from friction, air resistance, the elasticity of the string, etc. Again, these are deemed negligible with respect to the regularity that relates the period of the pendulum to the length of its string. Thus, they do not need to be included in the model. At least this is the case for a pendulum on Earth, but imagine that you were a physicist on a planet where the density of the atmosphere changes by several orders of magnitude on a regular and rapid basis. In this possible world, air resistance would be a central factor in explaining the period of a pendulum, perhaps far more important than the length of the string. So the notion of an "ideal" pendulum (or any other idealization) might depend on the environment in which the pendulum operates. Other good examples of abstraction are mechanical models of planetary motions, which abstract away from all properties of planets except for their relative sizes.

9.2.3 Distortion (Galilean Idealization)

The ideal gas law ($PV = nRT$) relates the volume V and pressure P of a gas to its temperature T (n and R stand for the number of moles and the gas constant, respectively). Clearly, if you increase the temperature and leave the volume alone, you increase the pressure of a gas. This relationship can be derived from the kinetic theory of gases, according to which gases are collections of molecules that move around rapidly and randomly (remember Maxwell's demon). Their motion translates into temperature manifesting mean molecular kinetic energy. Thus, if you increase the speed of the gas molecules, you increase the number of collisions between them and the walls of the container, thus increasing the pressure. However, the derivation of the law from the kinetic theory of gases requires some distortions, i.e., false claims about the gas. In particular, the gas molecules are treated as point masses whose collisions with each other and with the walls of the container are fully elastic, and which do not attract or repell each other.

In ecology, the population growth model began with simple, exponential growth: $dN/dT = rN$ (where r is the intrinsic rate of natural increase or the birth rate minus the death rate). The model abstracts away from all external factors. However, it also allows a population to grow to infinity! So modelers incorporated an environmental factor called *carrying capacity* (K) to put a limit on growth: $dN/dT = rN - (rN^2/K)$. Furthermore, the logistic growth model abstracts away from a population's response time to conditions. That is, the model assumes an instantaneous response of the population to its present size. But real populations often respond in a time-lagged manner. So the model can be modified as: $dN(t)/dT = rN(t) \{1 - [N(t-t_{lag})/K]\}$. The time-lagged response generates an oscillating, rather than stable or continuous, dynamic as the population approaches carrying capacity. But this improved logistic growth model still abstracts away from what might be an important consideration: living organisms move. The logistic growth model assumes that emigration (those moving out of the population) is equal to immigration (those moving into the population). As you can imagine, when the model includes movement, the population dynamics can change dramatically, and the mathematics involved becomes far more complicated.

In economics, we often assume that the agents making up the system we are trying to model are completely rational and fully informed. For example, typical models of international trade assume that all agents have perfect information about the value of the commodities, have rational preference rankings (in particular, their preferences are assumed to respect transitivity, meaning that if they prefer A to B and B to C, then they prefer A to C), and that the preferences of all agents can be aggregated. As some authors have pointed out, this feature makes economic models particularly vulnerable to the charge of being nontestable, for if the predictions of an economic model fail to pan out, the economist can always blame it on the irrationality of the actual agents. This is one of the ways in which distorting idealizations can be problematic for models.

9.3 Analogies

Scientists not only use models for exploring target systems, they also use analogies for reasoning about them. In this section, we briefly discuss the relation between models and analogies. To start with, models and analogies have something in common – they are both similar to their target systems. For models, we identified the relevant similarity as grounded in partial isomorphisms, which means that the similarities are purely structural. While analogies are also structurally similar to their target systems, they often differ from ordinary models in two ways. First, they might involve a shift from one set of concepts to an entirely different set of concepts. Second, the inhabitants of the structures that are isomorphic may have similar properties, both compositional and functional. A couple of examples should clarify these claims.

9.3.1 Conceptual Shifts

In the 1860s, James Clerk Maxwell proposed an analogy between the "flow" of electricity and the flow of a special liquid in order to better understand how certain phenomena, produced principally by his colleague Michael Faraday, could be produced in electrical systems. In this analogy, he represented the flow of electricity as the flow of an ideal (weightless, nonviscous, frictionless) fluid through a system of pipes, sources, and sinks. With this analogy, Maxwell was able to derive equations for the flow of electricity which generated Faraday's experimental data. In other words, the model was *empirically adequate* – what it said about the observable data was correct. However, Maxwell warned explicitly that one should not confuse his model (he called it an analogy) with a physical hypothesis about the nature of the flow of electricity. Electricity, he emphasized, is not an ideal fluid. However, it can be modeled as if it were: If electricity *were* a fluid with certain properties flowing through a plumbing system of pipes, sinks, and sources, it *would behave* in certain ways. In particular, it would behave in ways that generate the data we actually observe.

Notice the counterfactuals in the last couple of sentences. They indicate that electricity ought not to be taken as a fluid; it just behaves *as if* it were a fluid. Compare this analogy with the ideal gas model. Here, we postulate, in explaining the Boylean behavior of gases, that they consist of particles whose extensions are negligible and which don't exert any long-range forces onto each other. One proposed role of the introduction of such blatant falsehoods is to indicate that the actual truth of the matter doesn't make a difference in the production of Boylean behavior.[4] However, despite these falsehoods, they are still falsehoods *about the same types of entities*. The model still involves gas particles that collide with container walls and each other, as we think happens in the target system. In contrast, the fluid analogy departs from the target system in a more radical way – electricity is now conceptualized as a fluid, which it definitely is not. Clearly, the analogy involves a mapping from one conceptual system into another conceptual system, preserving certain structural features. For this reason, we may classify analogies as *metaphorical models*. What is the role of such models? For Maxwell, they are an aid to the understanding. They help us to grasp connections and regularities that are otherwise hard to understand. We will revisit the relations among analogies, metaphors, and understanding in the last section.

9.3.2 *Property Sharing*

A famous analogy is that between light and sound. Just like sound can be reflected and produce echoes, light can be reflected. Just like sound can be diffracted, light can be diffracted. Just like sound has an intensity, called loudness, light has an intensity, called brightness. Just like sound has a quality, called pitch, light has a quality, called color. It would seem that on this abstract level of description, sound and light share a lot of properties. Thus, one might infer that they share further properties: Just like sound is a wave propagated in a medium such as air, light was postulated to be a wave propagated in a medium, which we may call *ether*. In fact, Maxwell proposed models of the luminiferous ether (the light-bearing medium) in an attempt to derive the sound-like properties of light from properties of the ether. It is unclear whether he thought that the luminiferous ether was a mere analogy, or whether he proposed it as part of a conjectural model. Some of his contemporaries did take the claim that light needs a luminiferous ether at face value. For example, Dutch physicist Henrik Lorenz proposed that the so-called Lorenz contraction was a result of mechanical deformations of objects due to the resistance of the ether through which it travels, and American physicists Albert Michelson and Edward Morley conducted their famous experiments in an attempt to determine the speed of the earth relative to that ether, which led to the famous null-result that no such speed was detectable. At any rate, the structure of this sort of *analogical inference* seems to be this:

1. Sound and light share a number of observable properties.
2. Sound needs a medium for its propagation.
3. Thus, light needs a medium for its propagation.

Clearly, an analogical inference is quite risky, as shown by the fact that the conclusion of this one is false. In general, from the fact that two systems share a number of properties, it doesn't follow deductively that they share all of their properties.

A trio of cases from medical science illustrates the range of possibilities when it comes to analogical reasoning – from the evidently beneficial, to the probably harmless, to the potentially dangerous. In 1934, a pharmacologist made a fortuitous observation. After mice ingested meperidine, their tails took on an S-shaped curvature. This same peculiar response had been seen after the administration of morphine, so the scientist reasoned analogically that meperidine also might be an effective painkiller. He was right about this property, and we now have the benefits of Demerol (the familiar brand name for meperidine).

A disease occurring in tropical Africa and Asia, caused by tiny worms that infect the lymph nodes, resulted in grotesque swelling of the legs and genitals along with a thickening of the skin. This condition is called elephantiasis for the similarity of symptoms to the bulk and flesh of elephants – and it is this analogy that led to the Azande people of North-Central Africa to treat victims by making an incision and rubbing in ashes from the burnt flesh of an elephant's leg using a faulty line of reasoning akin to that which undergirds homeopathy (i.e., "like cures like"). The intervention is ineffective and probably harmless as ashes are an unlikely source of bacterial infection.

In the early 1970s, studies found a correlation between the consumption of artificial sweeteners and the development of bladder cancer in rats. Reasoning that the urinary system of laboratory animals was analogous to that of humans, scientists warned of the carcinogenic potential of artificial sweeteners and the Food & Drug Administration banned the use of cyclamate (other sweeteners such as saccharine and aspartame were also suspected). Subsequent research revealed that the original reasoning was flawed due to differences in the physiology rats and humans. The erroneous analogy generated significant industry losses and regulatory costs. Moreover, recent studies have found that some artificial sweetener derivatives actually interfere with tumor-associated enzymes.

On the other hand, analogical inferences can be used as heuristic devices, stimulating further research. As an example, consider the following analogy by Mary Hesse: Wings are to birds as fins to fish, and lungs are to birds as gills are to fish. Here we see a sharing of functional properties (lungs and gills have the function of supplying the organism with oxygen), and structural properties (wings and fins do not share precisely the same function, but are anatomically similar, propulsive appendages). Perhaps analogy points the scientist toward questions about what could ground it, and in this case, ultimately toward a broad evolutionary account of the development of life.

9.4 Learning from Models

The general inference enabling us to learn from models is a form of extrapolation: "What is true about the model is (approximately) true about its target system." Consider scale models, such as models of buildings used to test their capacity for withstanding earthquakes. Scientists or engineers build a model structure, the parts of which are appropriately scaled down, and then expose the structure to stresses and strains (also appropriately scaled down) to observe the behavior of the structure under the influence of earthquake-like forces. Finally, they extrapolate to what would happen to a corresponding real building during an earthquake – it either holds up to those forces with minor to major damage, or it gets destroyed. Of course, all of this is oversimplified, but it helps illustrate the general use of models in learning about real systems. Taking inspiration from R. I. G. Hughes[5], we can call this learning procedure **RDE**: *Representation – Demonstration – Extrapolation*.

- *Representation*: In our example, we designate a particular concrete structure as the representation of a real building. Notice that this not only involves scaling, but also abstraction: The color of the building, for example, is irrelevant to the modelling purposes.
- *Demonstration*: Next, we apply various forces to the model (in the case of the example, by making the surface on which it rests move horizontally and also bulge) to produce stresses and strains to observe what happens to the structure.
- *Extrapolation*: Finally, we infer that the behavior observed in the model under treatment would also be observed in the target system – i.e., a real building under stress. "What's true about the model is (approximately) true about its target system."

Clearly, learning from models fits our earlier characterization of inductive inferences as reasoning from the observed to the unobserved. We observe the behavior of a model, which we take to be a sufficiently good representation of its target system, and draw conclusions about the unobserved behavior of the target system. It comes as no surprise, then, that the process of learning from models is at least as risky as inductive inferences in general. We will look at some of those risks next, working our way up from the less risky (involving simple scale models) to the riskiest inferences from fictional models involving essential distortions.

9.4.1 Learning from Scale Models

Scale models are typically physical structures that resemble either larger or smaller real structures. Examples of the former include model buildings, bridges, cars, and airplanes, while examples of the latter include physically realized models of the hydrogen atom, DNA, cells, and various molecules. A special case of scale models is seen with one-to-one scale models, such as the plastic skeleton you might see in a physician's office.

The simplest cases of learning from such models are those in which the scaling of the observed model behavior, including the effects of that behavior, is linear. For example, it turns out that the drift of snow behind snow fences scales up linearly from 6-inch-tall models to 10-foot-tall real snow fences.

That is to say that the ratio of fence height to the extension of the drift in front of the fence is (nearly) constant from model to real system. Thus, we can find the best way to construct a snow fence by trying to optimize drift in the model, experimenting with different slopes, lattice spacing, etc., and then extrapolating the results to the real snow fence.

However, in many other cases, scaling is nonlinear. A famous example is the nonlinear scaling of muscle strength which is approximated by its power:mass ratio (the larger this value, the stronger the muscle). The power of a muscle is a function of its cross-section while the mass is a function of its volume. Consider a tiny muscle that we can estimate (as an elongated box) to be 1 mm × 1 mm × 10 mm (the size of a large grasshopper's femur). The power of the muscle would be 1 × 1, while the mass would be 1 × 1 × 10, so the power:mass ratio would be 1/10 = 0.1. Now, let's scale the muscle up to a human thigh which is 200 mm × 100 mm × 300 mm The power of this muscle would be 200 × 100, and the mass would be 200 × 100 × 300, so the power:mass ratio would be 20,000/6,000,000 = 0.003. So by virtue of nonlinear scaling (note that the power is a squared value and the mass is a cubed value), the grasshopper's strength is 30-fold (0.1/0.003) greater than that of the human.

Finally, there are additional problems when we scale only a part of a system. For example, suppose you want to test flight behavior by building a fixed-wing model of an airplane that's only half an inch long. You'll notice quickly that the thing doesn't fly at all. The reason is that at this scale, the viscosity of the air (which you didn't scale down) approaches that of a liquid, so that the Bernoulli effect, responsible for keeping planes airborne, doesn't take place anymore. In fact, insects do not so much fly through the air as they swim in it! Indeed, the movement of their wings is virtually identical to that of an oar rowing through water (their wings do not simply flap up-and-down but rotate while rising and falling). In this sort of case, we can speak of partial scaling: the behavior of the scale model is different from the real system, as the environment of the real system cannot be scaled down to an environment appropriate to the model.

9.4.2 Learning from Approximation Models

An approximation model is a model that incorporates a variety of abstractions, or Aristotelian idealizations. The ideal pendulum is such an approximation model. It abstracts away from friction and air resistance. Thus, its behavior approximates that of a physical pendulum. This can be seen by noticing that the curve describing the ideal pendulum approximates the more complicated curve we get from connecting the measured data points when we plot T against l in $\left(T = 2\pi\sqrt{l/g}\right)$. Of course, we can also look at this from the other perspective: The behavior of physical pendula approximates the behavior of the ideal pendulum, and the degree of approximation is measured by goodness-of-fit (i.e., sum of squares). Physical pendula are approximately ideal pendula. This was arguably Galileo's perspective.

Despite the presence of abstractions, the process of extrapolating from approximation models can be justified, if it is possible to de-idealize the model by introducing error-terms. For example, if we successively introduce terms for friction and air resistance, and we get increasingly better approximations to the behavior of physical pendula, we can conclude that in the limit of correct error-values, the theoretically predicted period-length function will fully coincide with the measured period-length function (the property of achieving ever better approximations by de-idealizing is sometimes called *monotonicity*).

Alternatively, we can try to "idealize" the physical pendulum by reducing the influence of air resistance and friction. While we can put the pendulum into a vacuum, we can't get rid of friction entirely. However, if the "idealized" physical pendulum approaches in its behavior that predicted by the ideal pendulum law, we have evidence for monotonicity again.

Likewise, population biologists have built chemostats which are chambers designed to study microorganisms in a constant environment. Fresh medium is continuously added and waste liquid is removed to keep the volume constant. So the chemical environment is unchanging or static (hence, "chemo" + "stat"). Of course, we can't eliminate all environmental constraints (the size of the chemostat can't expand indefinitely). But the "idealized" biological population dynamics resemble the logistic growth model. As such, it appears that we have a monotonic model.

9.4.3 Learning from Fiction (i.e., Distortion Models)

The epistemically most difficult cases involve models with Galilean idealizations, i.e., distortions. Examples include the ideal gas model which postulates infinitely many point particles, and economic models that involve rational and fully informed agents. How could we possibly learn something about the nature of the target system and its behavior from models that introduce obviously false assumptions? A similar question arises with respect to analogies. As we pointed out earlier, they are in a sense even further removed from their target systems by not just introducing falsehoods about the inhabitants of the target system, but by introducing entirely new kinds of inhabitants. How could such *metaphorical models* allow us to learn anything?

Perhaps a controversial answer to the last question can be given once we focus on different aspects of the process of learning from models. So far, we have posed the problem as one of how fiction can tell us anything about the nature of the real world. There is a gap between fiction and the real world that seems unbridgeable. However, it might just be the case that many aspects of the real world can only be understood by humans in terms of stories that are not literally true. Learning from distortion models might consist in the development of a capacity to grasp the real world in a metaphorical way, by mapping the conceptual structure appropriate for the real world (this conceptual structure is often mathematical in nature) onto a set of concepts that are more familiar to us. Take Maxwell's flow analogy. We have a much better grasp of how fluids behave in systems of pipes than we have of the propagation of electrical current as expressed by the relevant formalism, and it is not too difficult to see how ideal fluids would behave in such a system by abstracting away from properties of ordinary fluids, such as viscosity, temperature, etc. What seems to be happening is that we engage in a twofold mapping. First, we construct a mapping from one conceptual structure (the math describing electrical currents) to another conceptual structure (fluids) and then build an idealized model of that fluid by abstracting away from properties of ordinary fluids. Thus, we have an idealized model of a system that in turn stands in a metaphorical relation to the target system. The new structure involves inhabitants, and thus concepts, with which we are familiar. Maxwell's analogy was intended to provide us with a better grasp of what is happening.

Recent research in cognitive science has shown that analogical mappings, which often include conceptual shifts (achieved by introducing metaphors), have beneficial cognitive effects. Here is an example. Medical students are given the following problem: You have a patient with a large brain tumor that is located in the center of the brain. It has to be destroyed, which can only be done by radiation. However, the dosage required to destroy the tumor is so high that it would destroy far too much of the intermediate brain tissue, leaving the patient cognitively disabled, if not dead. What do you do?

After being informed of the problem, they are also given the following problem. You are the leader of a medieval army needing to sack a city that is located at the convergence of several rivers and in fact surrounded by those rivers. If you send in enough soldiers to take the city, any of the bridges over which you send the soldiers will collapse, and the city won't be taken. What do you do?

You may have a ready answer for the second scenario. Divide your soldiers into smaller groups and let them attack the city from three different directions. Now, do you see how to solve the tumor problem? To many, the answer becomes obvious: Divide the radiation into several beams that converge on the tumor. The net radiation arriving at the tumor is sufficient for destroying it, but each individual ray is weak enough so as to not damage too much of the intermediate tissue. The example shows that *analogical inference* can aid tremendously in problem solving. Thus, analogies, or what we called *metaphorical models* can help with reasoning about the real world. Radiation rays are like groups of soldiers; their convergence is like the soldiers meeting in the city to conquer it, and so on. While nobody claims that we learn something about the actual nature of the target system through such modelling, analogies can help us in reasoning about the target system. Thus, one straightforward answer to the question of how we can learn something useful from mere fictions is that we can learn how to reason about reality.

Perhaps there is a general lesson here. One might say that human understanding essentially involves narratives – either about human-like entities and their motives (think about mythological accounts of the origin of the universe), or more abstract causal narratives involving nonhuman agents, such as forces, motion, space, time, and what have you. Metaphorical models, such as Maxwell's flow analogy, are important as they present a narrative that leads to understanding: Electricity behaves as if it were a fluid flowing through systems of pipes, sources, and sinks. Even fictional models can thereby afford us a sense of understanding.

The second way in which we might be able to learn from fictional models is by looking at them as familiar instances of more abstract structural relations within the system we are trying to understand. On this view, even a fictional model can be correct about the system – not about the agents involved in the system, but about the relations among the agents (whose nature is irrelevant). And if the observed behavior is generated by certain structural features of the system, we get true understanding of the system on the basis of models that include false assumptions about the components of the system. This was Maxwell's claim: Faraday's data were generated by any system that is structured like a system in which an ideal fluid flows through pipes. Likewise, Richard Dawkins has described the evolutionary process in terms of the "selfish gene." Of course, he doesn't believe that genes have mental states that cause them to deliberately act in their own self-interest. Rather, this false assumption allows us to understand why biological systems behave as they do – they are structured as if greedy individuals (genes) were in control. On this view, the fictional model is not true about the system's components, but about their interactions – even fictional models can capture the relationships and/or interactions within a system, that make up their structure. This sort of view is called *structural realism*, and we will revisit it in Chapter 15.

Two final remarks on the notion of structural similarities: There are structural similarities that are merely topological, as is the case in the analogy between electrical currents and ideal fluids. By "merely topological" we mean that, for example, no similarities in causal powers between electrons and fluids are assumed. In other analogies, we suppose that there are some similarities in causal powers between the inhabitants of the isomorphic structures: Groups of soldiers and rays of radiation are causally similar in that considered individually, they don't destroy the paths to their target, but when converging at the target, they are powerful enough to "take" the target. Thus, some models, metaphorical ones included, make claims about the interactions and causal relations among the inhabitants of the target system. The nature of causal relations and interactions is the topic of the next chapter.

9.5 Conclusion

Before encountering the manifest forms of modeling in science, you could be forgiven for not seeing the common ground shared by subway maps, hydrogen atom diagrams, invisible screws, population dynamics, electrical flows, miniature snow fences, chemostats, and radiation treatments. But, of course, all of these are science fictions that tell us something about the workings of the natural world, whether through approximation, abstraction or distortion. Indeed, modelling is like scientific storytelling for grownups. Of course, this analogy is itself a model about modelling that might cause you to wonder about (or perhaps marvel at) the workings of science. And speaking of causing, that's the topic of the next chapter.

Notes

1   By © OpenStreetMap contributors – https://www.openstreetmap.org/#map= 15/48.2080/16.3663, CC BY-SA 2.0, https://commons.wikimedia.org/w/index. php?curid=70748821 selection/section.

2   By de:User:Häsk, correction de:User:Extrawurst – German Wikipedia: Bild:U-Bahn, S-Bahn, Wien.PNG, Public Domain, https://commons.wikimedia. org/w/index.php?curid=2369658.

3   Technically, this "definition" is incomplete, because it leaves out that the mapping can be inversed. This means that if the elements of A can be mapped onto the elements of B in a relation-preserving way, the elements of B can be mapped onto the elements of A in a relation-preserving way (just remember that if A is similar to B, then B is similar to A). However, because the intended function of a model will sort the two isomorphic structures into model and modelled, this complication is not important for our purposes. Since we intend, for example, to use the map to navigate the city, and not the other way around, the symmetry of the isomorphism between map and city becomes irrelevant.

4   See, for example, Michael Strevens, Depth. An Account of Scientific Explanation. Harvard University Press, 2008, p. 316ff.

5   "Models and Representation," Philosophy of Science, 1997, Vol. 64, No. Supplement, pp. S325–S336.

Annotated Bibliography

Daniela Bailer-Jones, 2009, *Scientific Models in Philosophy of Science*. University of Pittsburg Press,
   Pittsburgh, PA. The author discusses various types of models, including equations, animals, and physical scale models, and shows how they can be applied. She also provides an illuminating analysis of the relation between models and analogies, showing how analogies can sometimes guide model construction.

Roman Frigg, 2020, "Models in Science," Stanford Encyclopedia of Philosophy, available at https://plato.stanford.edu/entries/models-science/.
   An advanced overview of the epistemology and metaphysics of models, providing additional detail about various discussions of models in the literature.

Isabelle F. Peschard and Bas C. van Fraassen (eds.), 2018, *The Experimental Side of Modeling*. University of Minnesota Press,
   Minneapolis, MN. An excellent collection of papers discussing the practice of modelling in disciplines such as high energy physics, astronomy, ecology, fluid dynamics, cognitive science, the social sciences, and synthetic biology, among others. The editors' introduction presents a useful history of the changing attitudes toward modelling in the philosophy of science.

10

CAUSATION AND CAUSAL INFERENCE

10.1 What's the Problem with Causation?

Many scientific theories claim to reveal causal relations between events: gravitational attraction causes unsupported objects to fall to the ground; wolves are causally responsible for the death of livestock; the latest unemployment numbers cause the stock market to tank; and so on. Of course, causal talk is not confined to the sciences – we engage in it all the time when we talk about why certain things happened: a wayward baseball caused the window to break or my car troubles caused me to be late. Causation is so deeply entangled with our attempts to understand the world that some philosophers of science have argued that good explanations – the principal means for gaining understanding – typically identify causes of the events we try to understand. Clearly, then, it is important to have a good grasp of the nature of causal relations.

At first glance, attaining such a grasp seems easy enough. Just observe the events that led up to the event you want to understand and voila – you discover its cause. On this naïve view, causal relations are simply relations between one event and another event that follows it. However, a brief moment's reflection shows that this can't be right: Not all events that are prior to some event e in which I am interested are causes of e. Suppose that just before the baseball crashes through my kitchen window, a tile drops from my bathroom ceiling. Clearly, the tile's dropping from the ceiling didn't break my window, despite the fact that it occurred just before the window broke. The tile is simply too far away from it. This suggests that some event e^* causes another event e just in case that e^* is both prior and sufficiently close to e.

This is Philosophy of Science: An Introduction, First Edition. Franz-Peter Griesmaier and Jeffrey A. Lockwood.
© 2022 John Wiley & Sons, Inc. Published 2022 by John Wiley & Sons, Inc.

And since we can observe both temporal sequence and spatial distance, we can observe that e^* causes e. Good news for the empirical sciences, whose bread and butter is observation!

Unfortunately, things are not that easy. There are examples of events that stand in the right spatiotemporal relation to other events without being the causes of such events. David Hume was perhaps the first one to make this possibility explicit. In the next section, we will discuss his claim that the right spatiotemporal relations are not sufficient for grounding causal relations. In a nutshell, what Hume revealed was the fact that the concept of causation is a *modal concept*, and thus involves relations between events that cannot be directly observed. If Hume's right, this spells trouble for empirical theorists who want to make causal claims, because it is now unclear how such causal claims can be justified on the basis of observation alone. More recently, Wesley Salmon and others have tried to replace the modal condition with a non-modal condition in their so-called *conserved quantity theory* of causation. For an empiricist, such a theory looks initially promising. However, it is of dubious applicability to any causation other than physical causation, as we will see.

10.2 Hume's Challenge

There are plenty of cases in which people confuse a mere correlation with genuine causation – this is the general problem of confounds, which we have already discussed in Chapter 6. For example, an observed correlation between taking a pill and improving one's health is not a strong enough reason for inferring a causal relation between the pill and improved health – in this case embodied in the claim that the pill is effective, or causally efficacious. We usually need to rule out confounds, by using both negative and positive controls. The deep reason for this is the simple fact that causation itself seems unobservable. Were it observable, we would not have to wonder whether or not some effect could really be attributed to a specific agent, such as the pill. We would simply be able to observe the pill causing the improvement in health. But alas, that doesn't seem to be possible, so we are forced to attempt to triangulate the causal factors by using treatment and controls.

The reason for the unobservability of causation was first clearly elucidated by David Hume. He was a strict empiricist, meaning that he subscribed to the doctrine that all knowledge has to be ultimately grounded in observation. He argued that since causation is not observable, we cannot have any causal knowledge. But why did he think that causation is unobservable? Here is a modern reconstruction of his reasoning.

Hume analyzed our ordinary notion of causation by comparing cases in which we are willing to attribute causation with those in which we refuse to do so. We have already seen examples of such cases. The tile falling from my bathroom ceiling couldn't have been the cause of the breaking of my window, because the tile wasn't anywhere near my window – but the baseball was. Moreover, the baseball hit the window just before it broke; on the other hand, my angry cursing at the neighborhood kids occurred just afterward. So, the baseball seems to be the best candidate for a cause, as the tile is out for reasons of being too far away and my cursing because it came after the window's breaking. This suggests two necessary conditions that any causal relation between events has to satisfy: a *locality condition* and a *future-directedness condition*. The question is of course whether those two conditions are also sufficient. In other words, are all events e^* that are prior to some other event e and are located in the same region as e causes of event e? The answer seems to be a resounding "no": Imagine that just before your window breaks, you very gently touch it. A fraction of a second later it breaks. Did your touch break the window? It wouldn't seem so. But if it didn't, then being temporally prior and spatially close to some event e is not sufficient for being the cause of e. Something is missing.

What may come to mind at this point is that your gentle touch isn't strong enough to break the window, but also that had you not touched it, it still would have broken on account of the baseball. The second thought clearly involves a counterfactual, and we'll return to it in a minute. What about the first thought, that your gentle touch couldn't have broken the window? How do you know that?

The obvious answer is: from experience. You have data attesting to the fact that gentle touching doesn't break windows. You have observed it over and over again – touch a window gently, and it doesn't break. Setting aside the whole problem of induction, what you have observed is a correlation between a lack of contact-force and a lack of window-breaking. Moreover, it seems clear – especially in light of the baseball that hit the window at high speed – that a stronger touch (like the baseball's) would break the window. Thus, the idea that the lack of force is responsible, in cases in which you touch the window gently, for the fact that it doesn't break is supported by another counterfactual: Had you touched the window with greater force, it would have broken.

It seems, then, that identifying causes for events is based on counterfactual reasoning: Had the baseball not struck the window, it would not have broken, despite the fact that I gently touched it. Thus, in claiming that event e^* caused event e, we seem to claim that if e^* had not occurred, e would not have occurred either. Recall, though, that the truth conditions for counterfactuals involve other possible worlds. A counterfactual is true if and only if in all the nearby possible worlds in which the antecedent is true, the consequent is true as well.[1] As we have noted before, we cannot observe with our senses what happens in other possible worlds. Putting aside for the moment the difficult problem of how we can know whether a counterfactual is true or not (if indeed we ever know such things), we can see why this modal condition on causation upset an empiricist like Hume.

10.3 Causation as Mere Regularities

As a strict empiricist, Hume subscribed to the doctrine that any empirical knowledge about the world must ultimately be grounded in observation. Clearly, any claim that some event e^* causes some other event e (such as the event of the baseball striking the window causes the event of the window breaking) is an empirical claim about the world. In order to be knowledge, it has to be grounded in observation. However, the above analysis shows that causal claims involve counterfactuals, the truth of which cannot be determined by observation. Therefore, strict empiricism implies that we cannot have causal knowledge. Hume accepted this consequence and proposed a *psychological account* of our causal beliefs. If we form, on the basis of observation, the belief that some event e^* is happening, and if we have seen e^* correlated with e in the past, our mind simply has the tendency to automatically form the belief that e will happen as well. This sort of automatic association between our beliefs about e^* and e is what we mistake for an objective causal relation between the events. Causation is a kind of associative learning – it's all just in the head (think about Pavlov's dog which salivated when a bell rang because the ringing was systematically arranged to precede food)!

In fact, causation becomes a kind of superstition. The famous psychologist B. F. Skinner conducted a study on the formation of superstition in pigeons. He attached to the birds' cages a device that automatically provided food at random intervals – no matter what the pigeon was doing. The birds connected the arrival of a treat with whatever action that they had been performing by mere chance when the pellets dropped into their food bowls. Associating their actions with the arrival of food, the pigeons repeated their behavior with dedicated intensity – and, sure enough, another allotment of food would appear if they just persisted long enough! One bird associated turning counterclockwise in its cage with the arrival of chow, while another became convinced that screwing its beak into a corner of the cage did the trick. Hume would've considered these pigeons much like humans who mistake some event as causing another.

Some contemporary empiricists do not like Hume's solution. It just seems that there is a difference between mere psychological association and genuine causation. Somebody might strongly associate the sight of a black cat with impending bad luck (and might even be reasonable in doing so, if he indeed reliably experienced bad luck after seeing a black cat in the past), but we won't take this as sufficient grounds for claiming that black cats actually cause bad luck. But how can we find a difference between mere association and genuine causation without invoking the empiricistically dubious modal criterion?

10.4 Conserved Quantities to the Rescue?

Starting in the 1980s, Wesley Salmon developed, along with other similarly minded empiricists, a theory of causation that allows us to rule out spurious associations without invoking a modal condition. This account has become known as the *conserved quantity theory* of causation (CQT). The guiding idea is simple. *Causal* interactions between objects are those interactions that *change* some aspects of the behavior of both objects in such a way that the change *persists* even after the interaction takes place. The window stays broken after the baseball strikes it, and the baseball's velocity remains below its original level after it leaves the shattered window behind. To formulate the account more precisely, CQT assumes that objects can be represented by

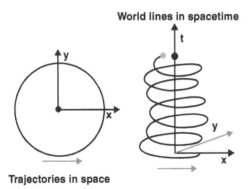

Figure 10.1 World lines.

their so-called *world lines*, which trace their motion through space and time, or space time. Consider a simplified representation of the earth's (dot) trajectory around the sun (x), disregarding the fact that the trajectory is elliptical, and disregarding one of the spatial dimensions, as shown in Figure 10.1:

In the right picture, Earth is represented as moving through both space and time. Metaphysically, this representation is often associated with the view that ordinary physical objects are in fact four-dimensional: They are extended in three spatial dimensions and in one temporal dimension. While on more traditional views, a three-dimensional object is fully present at any moment in time at which it exists at all, on this view, only a time-slice of a four-dimensional object is fully present at any given moment of time. Traditionally, you fully exist at 3:00 P.M.on 3/31/2022, while on this view, only a part of you exists at 3:00 P.M. on 3/31/2022. We will encounter world lines in our discussion of the theory of relativity as well.

Given the conception of objects as four-dimensional entities, we can represent their interactions as intersections of world lines. (We count "touching" as an intersection for our purposes.) If during such an intersection, a change happens in both world lines that persists after the intersection, we talk about a causal interaction. Now consider a game of pool. You hit the cue-ball, it hurtles toward the 8-ball, the two collide, and the 8-ball rolls toward the corner pocket, while the cue-ball moves toward the middle of the table. In our new language, we describe it as follows: The cue-line approaches the 8-line, the two touch, and afterward, the 8-line and the cue-line separate, both moving through time and space. Physically speaking, we witnessed the transfer of some of the linear momentum from the cue-ball to the 8-ball. Thus, during the intersection of the two world lines, linear momentum was exchanged. Linear momentum is (ideally) a conserved quantity. Thus, the proposal is that causal interactions are those intersections of world lines during which a conserved quantity is exchanged. This is CQT.

10.4.1 Can CQT Deliver on All Fronts?

Salmon originally proposed CQT in the context of developing a model of scientific explanation according to which all such explanations are causal in nature. It turns out though that many intersections of world lines can involve exchanges of *multiple* conserved quantities. For example, during the collisions of billiard balls, not only linear momentum is exchanged. If the balls are made from the right material, electrical charge, another conserved quantity, is also exchanged. If I want to explain why the 8-ball ended up in the corner pocket, I will, however, not invoke the charge exchange, but the momentum exchange instead. But there is nothing in CQT that tells me which of the exchanged quantities are relevant for explaining the change in movement. An extra story about relevance is needed here.

However, apart from this problem of connecting CQT with explanations, the theory looks like a success for empiricists. Both the intersection of world lines and the exchange of conserved quantities can in principle be observed. In effect, CQT strips the notion of causation of its modal involvement. No talk of counterfactuals, necessary connections, and the like. All seems to be on the up-and-up. Nevertheless, there is an important question raised by CQT: Does it propose an account of causation that is applicable to every causal claim we might make in the sciences? For example, does biological causation involve the exchange of conserved quantities? *If* reductionism is correct, and we can reduce all biological regularities to physical regularities, then it might work. However, there is also the view, discussed in Chapter 14, that such reductions are impossible. In that case, biological causation, such as when certain hormonal deficiencies cause mood changes, cannot be captured by CQT. Other problematic cases involve causal explanations in the social and behavioral sciences.

A final problem concerns so-called *causation by omission*. If I fail to water my garden plants for long enough, I cause them to die. It is highly obscure, to put it mildly, how this case of causation could be spelled out in terms acceptable to CQT. What could possibly be the conserved quantity exchanged between my failure to water the plants and their demise? Faced with examples of this kind, some have argued that perhaps our notion of causation is a cluster concept: Causation in physics is one thing, while causation in gardening is quite different. On this view, CQT provides a reductive analysis of causation as it is used in physics, while it is simply not applicable to phenomena that fall outside the domain of physics. This move raises the question of why we call both the interaction between the balls on a billiard table and the lack of interaction between gardeners and plants examples of causation. All things being equal, a unified account of what causation is would be preferable.

10.5 Causation and Manipulation

One attempt to develop such a unified account of causation that is applicable, not only in physics, but also in the biological, social, and behavioral sciences, is called the *manipulationist* account. The main idea is simple: X causes Y just in case that I can manipulate Y by manipulating X. Recall the billiard balls: I can manipulate the trajectory of the 8-ball by manipulating the trajectory of the cue-ball. Similarly, you can manipulate the life and death of my plants by manipulating me – inducing me to water them by offering me some money or, alternatively, restraining me so that I can't water them. In general, if we want to find out whether some X is causally connected to some Y, we can try to manipulate X and see whether such manipulations result in systematic changes in Y. In Chapter 3, we discussed drug trials, which are of course attempts to establish the existence of relevant causal connections. The manipulationist approach fits such trials perfectly: By manipulating whether or not a person gets the drug, we hope to show that we can manipulate whether or not that person recovers from the disease.

A further advantage of this approach is the ease and naturalness with which we can distinguish between mere correlation and causation through manipulations. Barometer readings are reliably correlated with certain changes in the ambient weather. A falling barometer reading means that a storm is approaching. However, the barometer reading does not cause the storm to approach. Rather, both the storm and the reading are the results of a common cause: falling air pressure. That the barometer doesn't bring about the storm can easily be shown through a manipulation. If we *intervene* and set the barometer reading manually, in a way that is statistically independent from changes in air pressure (by using a random number generator, say), we will see that such a manipulation does not allow us to control the emergence of a storm. Thus, even when two events are highly correlated, we should only infer a causal relation between the events if intervening with one provides control over the other.

In addition, the manipulationist approach provides a clear distinction between direct, contributing, and total causes. Consider the billiard example once more. When I strike the cue-ball with my cue, and the ball later collides with the 8-ball, sending it toward the corner pocket, my striking the cue-ball is not a direct cause of the movement of the 8-ball. Someone could intervene and stop the cue-ball before it collides with the 8-ball. Thus, X is a *direct cause* of Y if there are no possible interventions that change Y other than interventions on either X or Y. The collision between the cue-ball and the 8-ball sends the latter toward the corner pocket, and in order to change that fact, I would have to either prevent the collision (X) or stop the eight ball as soon as the collision happens (Y). Thus, the collision is a direct cause.

Return now to the plant example. If my neglect to water the plants kills them, my watering the plants causes them to stay alive. Of course, watering them is not a sufficient cause. Among other things, the plants need access to phosphorus, which they can get either naturally from the soil or through addition of fertilizer. However, if we hold access to nutrition fixed, my watering the plants will cause them to stay alive. Thus, watering is a *contributing cause* of their continued survival. This has the consequence that, for example, by removing access to nutrients, the plants can be caused to die.

Finally, we can define X as a *total cause* of Y if there are at least some interventions on X that make a nonzero difference for Y. In this sense, I can be a total (but not full) cause of my plants' survival, because someone can intervene and make me water my plants, which, if we keep access to nutrition fixed, survive as a consequence of this intervention. The manipulationist literature has developed a sophisticated machinery to analyze complex causal interactions, which are combinations of direct, total, and contributing causes as well as various combinations of interventions, by using directed graphs and structural equations. This machinery allows us to trace causal connections in many economic, social, and biological contexts. Exploring it further, however, is beyond the scope of this book.

However, it is worth pointing out that the manipulationist framework does not entail a locality condition for causation. As long as an intervention on X lets us control the behavior of Y, X is causally connected to Y, even if it happens to be the case that X is not spatiotemporally connected to Y. This feature of the manipulationist framework might be suitable for understanding some of the strange phenomena we encounter in quantum mechanics, such as those emerging from the so-called EPR thought experiment (to be discussed in the next chapter). Moreover, this lack of commitment to locality allows us to identify causal connections in cases in which the exact processes underlying the causation remain obscure. For example, television commercials can arguably be a contributing cause of somebody's buying behavior. Sure, there is a spatiotemporal process respecting locality that connects the commercial with its viewer. But what exactly goes on afterward? How does locality-respecting causation lead from viewing, through motivation, to eventual behavior? The manipulationist framework abstracts away from those details and concentrates on issues of intervention and control instead, which are often empirically ascertainable. Often, but not always.

10.5.1 Manipulation and Counterfactuals

As stated earlier, the cause of a falling barometer, as well as of the approaching storm, is falling air pressure. To show this, we would have to intervene in the air pressure and thereby control the emergence of storms. Well, we can't really do that, and there are numerous other causal claims in the sciences where intervention is out of the question. Just think about the cosmologist's claim that the universe came into existence as the result of the Big Bang. Or that the earth's orbit "around" the sun is really the effect of the sun causing a particular curvature of spacetime in its vicinity (as we'll see in the next chapter). In none of these cases are intervention and control possible. For this reason, the manipulationist approach has to resort to counterfactuals: If we *were* to appropriately intervene in the air pressure, a storm *would* be the result.

In a way, the reappearance of counterfactuals shouldn't come as a surprise. Just like the straightforward counterfactual analysis of causation, the manipulationist framework ties causation to facts about *dependence*. In the counterfactual analysis, this dependence is simple counterfactual dependence: A causes B just in case if A hadn't happened, B wouldn't have happened. The manipulationist approach claims that A causes B just in case the occurrence of B depends on the occurrence of A. This dependence is expressed in terms of interventions and controls. However, it is simply not possible to arrange for the appropriate interventions in all cases or even most of them. Thus, the dependence must be expressed in terms of counterfactual interventions and controls: Falling air pressure causes a storm just in case that in a possible world in which we intervene with the air pressure, we control the storm.

Put this way, the approach starts to look terribly anthropogenic. It seems to make causation dependent on what we can do, even if only counterfactually. However, this worry can be alleviated by thinking of interventions in a more abstract way that does not require human agency. To wit: An intervention on X is any exogenous cause that changes X. Going back to the storm example, we first identify a causal structure, consisting of the air pressure A, the presence or absence of storms S, and the barometer reading B. We can observe certain correlations between A, B, and S. In order to figure out the direction of the causal arrows in this structure, we need to wait for exogenously caused changes in any one of A, B, or S, and observe how the remaining factors behave. If a cold front (a cause exogenous to our system) changes A and thereby both B and S, we know that a causal arrow emerges from A. If, on the other hand, a random number generator (another exogenous cause) sets various readings of B, and we observe no systematically connected changes in either A or S, we know that there is no causal arrow emerging from B.

The astute reader will loudly protest at this point. Didn't we just use the notion of causation in order to provide an account of causation? This would indeed be a devastating objection, if the manipulationist framework were intended to provide an analysis of causation in noncausal terms, the way regularity theories, CQT, and the counterfactual analysis attempt to do. But that's not the project. Rather, taking causation as an unanalyzed term, manipulationists try to provide methods for analyzing the various dependencies in empirically discoverable causal structures. They do this by distinguishing between endogenous causation, such as the causation of storms by falling air pressures, and exogenous causation, such as a cold weather front intervening in the air pressure. No attempt at providing a theory of the *nature* of causation is made. However, manipulationists think that no such theory is needed for understanding one of the central applications of our notion of causation in science, viz., causal explanations. To understand the force of the latter, all we need is a grasp of interventions and controls and the resulting dependencies.

10.5.1.1 The New Mechanism

The manipulationist account can also be seen as a particular form of a movement in the philosophy of science known as "the new mechanism." Replacing the traditional emphasis on the relation of laws to phenomena, so characteristic of research in the positivist framework, new mechanists are interested in finding the mechanisms that underlie the production of various phenomena. Mechanisms don't have to be machines – there are biological mechanisms, such as photosynthesis, cognitive mechanisms, such as those involved in visual information processing, and economic mechanisms, such as those responsible for fluctuations in commodity prices, to name just a few. The "new mechanism" movement started with the realization that a law-centered approach to explanation (see Chapter 13) and other topics was not applicable in the life or social sciences. Thus, instead of explaining biological phenomena in terms of subsumption under biological laws (of which none could be found), philosophers argued that such explanations aim at identifying and analyzing the mechanisms that bring about such phenomena. We will have an opportunity to see a particular form of the new mechanism approach in action when we briefly explore the nature of cognition, as well as the difference between reduction and emergence, in Chapter 14.

10.6 Conclusion

We have encountered various reactions to Hume's challenge for causation. His own regularity theory is acceptable for the empiricist, but doesn't have the resources to distinguish between mere correlation and genuine causation. The conserved quantity theory, which provides an empiricist account of the nature of causation, is of dubious applicability outside the realm of physics. And finally, the manipulationist approach reintroduces some elements of the counterfactual approach without pretending to provide an analysis of the nature of causation. At best, it delivers powerful techniques for exploring the dependencies in given causal structures. Thus, we are still without a compelling story about what causation actually is. To make matters worse, we encounter pressures from the empirical sciences themselves to relinquish the locality and future-directedness conditions on our ordinary notion of causation. This is the topic of the next chapter.

Note

1 Those readers with some knowledge of formal logic might notice that a counterfactual is trivially true if its antecedent is impossible. In this case, the antecedent is false in all possible worlds, which in turn makes the conditional true (remember, conditionals with a false antecendent are true).

Annotated Bibliography

David Hume, 1748, *An Enquiry concerning Human Understanding*, edited by Tom L. Beauchamp. Oxford University Press, Oxford, UK, 1999.

Section 4 contains the famous criticism of the notion of causation from an empiricist point of view. In section 5, Hume develops his "skeptical solution" to the problem in terms of custom, or habitual associations between events.

Wesley Salmon, 1984, *Scientific Explanation and the Causal Structure of the World*. Princeton University Press, Princeton, NJ.

In this ground-breaking work, Salmon introduces the conserved quantity theory of causation. In distinguishing between pseudoprocesses and causal processes in terms of the capacity for transmitting marks, he argues that causal interactions are those in which marks, such as conserved quantities, are exchanged between causal processes.

James Woodward, 2003, *Making Things Happen. A Theory of Causal Explanation*. Oxford University Press, Oxford, UK.

A sustained defense of the manipulationist account of causation and how it can illuminate a number of questions about scientific explanations.

11

STRANGE CAUSATION – TIME TRAVEL AND REMOTE ACTION

In this chapter, we discuss how century-old developments in physics challenge both the assumption that causation must be local and that it must be future-directed. These developments are the results of the two central revolutions in physics in the early twentieth century – the advent of *Relativity Theory* and of *Quantum Mechanics*. In particular, Relativity Theory, with its revolutionary impact on our understanding of space and time, opens the possibility for backward causation (a cause *follows* an effect!) as a result of traveling into the past. These possibilities lead to surprising consequences for many familiar metaphysical assumptions. Quantum Mechanics, on the other hand, challenges the deeply engrained view that all causation is local. Equally surprising as backward causation from a metaphysical point of view, there is good reason for believing that causal interactions can occur between events that are separated by large spatial distances.

A warning before we get started: After reading this chapter, you should not believe that you now have a solid understanding of either relativity theory or quantum mechanics. We are only able to offer the most basic ideas behind those theories in highly simplified form. We don't think that our presentation contains serious distortions, but if you really, really want to understand either of those theories, you need do the work: Take advanced courses in physics, or at least study closely the suggested readings found at the end of this chapter. The best we can hope for here is to whet your appetite for a serious engagement with those theories by showing how they challenge some of our most deep-seated assumptions about the world.

This is Philosophy of Science: An Introduction, First Edition. Franz-Peter Griesmaier and Jeffrey A. Lockwood.
© 2022 John Wiley & Sons, Inc. Published 2022 by John Wiley & Sons, Inc.

11.1 On Influencing the Past

In the last chapter we encountered the traditional analysis of causation, which identified three conditions that causal interactions have to satisfy:

1. **locality**: there is no causation at a distance
2. **future directedness**: effects occur *after* their causes
3. **modality**: cause and effect are connected by necessity

We saw that so-called conserved quantity theories of causation attempt to dispense with condition 3 (modality) mostly for empiricist reasons. Instead, conserved quantity theories hold that causal interactions are those intersections of world lines during which conserved quantities are exchanged. As we pointed out, whether this solution works for all causal claims remains to be seen.

The other two conditions have been challenged as well, but this time due to developments in the empirical sciences themselves. In particular, the special and general theories of relativity challenge our strong belief that backward causation is impossible, while quantum mechanics throws into doubt the idea that all causation is local. We will start by briefly introducing, in this section, the fundamental ideas behind both the special and general theories of relativity and then discuss their consequences for the metaphysics of causation. In the next section, we will turn our attention to quantum mechanics and explore its implications for the claim that causation has to be local.

Before we get started, a couple of remarks on *metaphysics* are in order. Metaphysics has had a bad reputation for a while, either based on a thorough misunderstanding of what it is, or on a misconception of how it is related to empirical research. Historically, "metaphysics" is simply a librarian's term. When Aristotle's students compiled his diverse manuscripts, they put his work on abstract topics next to his writings on physics, *meta*-physics (just after, physics). Thus the name.

These days, you sometimes encounter "metaphysics" as a term used to elicit a feeling of transcendence and spirituality; it's often associated with "New Age" style approaches to the world. But angels, crystals, and tarot cards have NOTHING to do with metaphysics properly understood. Rather, metaphysics is concerned with the fundamental nature of reality and is based on the sort of reasoning that respects evidence and cogent argumentation as encountered in the various sciences and in philosophy. Despite their abstract nature, metaphysical questions arise in the center of our empirical sciences, even if, for legitimate or illegitimate reasons, scientists sometimes choose to ignore them. We have already discussed the relation between controls in experiments and counterfactuals, where the latter involve truth conditions defined over possible worlds. This raises the question of what the metaphysical status is of such merely possible worlds. Another question pertains to the metaphysics of mathematical objects: Are they real things, but abstract rather than concrete? Or are they just ideas in someone's head?

The nature of causation is another metaphysical topic. While our ordinary notion of causation seems to involve both locality and future-directedness conditions, developments in physics might force us to abandon those conditions. This means that metaphysics might not be just an a priori discipline (one that can make progress without any data, such as discovering the principles of logic), but might well need input from the empirical sciences. On such a view, metaphysics and physics lie on a continuum: Physics typically relies more on observation than on a priori reasoning, while metaphysics reverses this relation and uses more a priori reasoning than observation. However, both disciplines use both sources of knowledge. In a slogan, *metaphysics without physics is empty* (because it's purely speculative), *while physics without metaphysics is blind* (because it lacks conceptual clarity). The discussion that follows can thus be seen as pursuing metaphysics by paying close attention to physics itself: What general consequences for some of our most fundamental beliefs about the structure of the world do those theories have?

11.1.1 Basics of the Special Theory of Relativity (STR)

Some of the most surprising consequences of STR pertain to our everyday notions of space and time, or at least to our notions of space and time insofar as they reflect ordinary classical mechanics. In a nutshell, spatial distance and temporal duration turn out to be a function of motion, as is the simultaneity of events. These consequences can be shown rather effortlessly to follow from the two fundamental assumptions of special relativity, to which we now turn.

The first assumption is called the *principle of relativity* and says that if two bodies move inertially (i.e., not influenced by any forces) relative to each other, each of the bodies is equally correct in "ascribing" the relative motion to the other object. Arguably, this principle goes back to at least Galileo. Borrowing an image from the American physicist Brian Greene, if you were floating in outer space (protected by a space suit) and saw your buddy approach you and wave as she flew by, it would be equally correct for her to describe the situation as one in which she is just floating and you are flying by (in the opposite direction, of course). In more formal terms, the principle says this: *If an object moves inertially, then there is no experiment that could be used to determine whether the object is in motion or at rest.* Just think about taking a flight on an airplane during good weather. If there are no turbulences, and if the plane is at constant cruising speed, everything works as if you were at rest: You can pour coffee without spilling it, for example.

The second assumption is called the *principle of the constancy and maximality of the speed of light.* It simply says that no matter how fast an object moves, light recedes from it at the same constant speed. In other words, if you were to run with a flashlight that's turned on, then, no matter how fast you run, the light always leaves the flashlight at the same speed – you cannot catch up with it. The speed of light is independent from motion. More technically, *the speed of light is the same in all inertial frames.*

These two principles are sufficient to show something very surprising about time itself: Time slows down as a function of motion. More precisely, time elapses more slowly in a moving object as compared to a stationary object from which it is measured: If two objects, A and B, move inertially relative to each other, A will measure B's time to elapse more slowly, and B will measure A's time to elapse more slowly. Consider Figure 11.1, in which a moving light clock (right) is compared with a stationary light clock (left):

Both light clocks consist of perfectly parallel mirrors between which a single photon is bouncing up and down (without loss of energy). Since the speed of light is the same in both cases, but the light has to travel a longer distance on the right, the photon's roundtrip there will take longer than that of the left photon. If X roundtrips on the left equal one second, then the right photon will make Y < X roundtrips during that time, measuring time as elapsing more slowly. This *time dilation*, as it is called, is not just a theoretical result – it can be observed to *really* happen. Elementary particles called *muons* disintegrate, when at rest, in about two millionths of a second. However, if they are accelerated to a speed of 99.5% of the speed of light, their "lifetime" increases by a factor of about ten. If people managed to zip around

Figure 11.1 Light clocks.

at such high speeds, they would on average live (relative to the rest of us), not just 70 years, but 700! As we will see below, something remarkably similar happens to length – objects get shorter in the direction of their motion, as measured from another object relative to which they are moving (this is called length contraction). Before we get there, let's have a look at one startling consequence of time dilation.

11.1.2 When One Twin Is MUCH Older than the Other

Bob and Jim are twins, born just minutes apart in 1988. Bob, the younger of the two, joined the Air Force to become a fighter pilot, while Jim threw himself into an academic career, which earned him both an advanced degree in mechanical engineering and a PhD in theoretical physics. Over the last couple of years, he spent his summers with Bob working on plans for a new, extremely efficient and powerful rocket engine. In the winter of 2017, those plans were finally realized through a cooperation between NASA and CalTech, and in the spring of 2018, both Jim and Bob arrived at Cape Canaveral. Bob, now a trained astronaut, had been designated as the commander for a mission which would take him on a 5-year long trip to a neighboring galaxy. This had become possible for the first time in history, as the new craft was able to reach speeds of up to 80% of c (the speed of light). Jim stayed behind with the ground crew monitoring the journey. On May 25, 2018 – exactly on their 30th birthday – the spacecraft was finally launched. While not all details of the journey have been declassified yet, one thing has become abundantly clear upon Bob's return in 2023: Special relativity is correct! How did it become clear?

When Bob climbed out of the spacecraft, he looked remarkably young. As his brother Jim hugged him, the difference was quite startling – a balding and graying Jim stood next to a Bob with a full head of brown hair. Checking the computer clock on the spacecraft supported what STR predicted. In comparison with the clocks at mission control, the traveling clock had slowed down considerably. While for Jim, 5 years had passed, Bob's trip lasted for only 1 year, according to the traveling clock. Thus, Bob was now 31, while Jim, his twin, was 35! Bob had travelled into Jim's future.

11.1.3 Basics of the General Theory of Relativity

One of the basic assumptions of general relativity is the so-called *equivalence principle*. It says that the effects of gravitation are observationally indistinguishable from the effects of acceleration. Consider the following thought experiment. You are in a windowless elevator cabin. All of a sudden the elevator cable snaps, and the cabin falls freely toward the ground. What happens to you? If the acceleration is just right, you will float "weightlessly" in the middle of the cabin. Now compare this situation to one in which some evil demon turns on a strong gravitational field above the cabin (by, say, dragging the moon to a suitable distance from the top of the elevator). Again, you will find yourself floating "weightlessly" in the middle of the cabin. Given that there are no windows, you don't know, on the basis of observing yourself floating in the middle of the cabin alone, whether the cable snapped or a gravitational field was turned on. Observationally, the two possibilities – floating due to acceleration and floating due to gravitation – are indistinguishable. This is exactly what the principle says: Any effect we ascribe to acceleration can with equal justification be ascribed to gravitation.

There is a second phenomenon postulated by special relativity that we have already mentioned, but without much explanation. This phenomenon is called *length contraction* and is instantiated by moving objects. Roughly, the length of a moving object shrinks in the direction of the object's motion. Length contraction follows rather quickly from time dilation. Remember that we can measure the length of an object, or the distance between two objects, in terms of the time light needs to get from one (end of the) object to the other. For example, the distance from the earth to the sun is approximately 8 light minutes. Now imagine we measure the length of objects in terms of light nanoseconds, one of which corresponds to about 1 foot. Imagine further that two observers measure the length of two objects that are moving relative to each other. Given the principle of the constancy of the speed of light, and the phenomenon of time dilation, the moving object must have shrunk relative to the stationary object. To see this, remember that $v = d/t$ (velocity is distance divided by time). Clearly, if the speed of light measured by the person on the moving object, where time elapses more slowly, is to agree with the speed measured on the stationary object (which it must, given the constancy of the speed of light), the value of d for the moving object must be smaller than the corresponding value of the stationary object.

Now we have all the pieces required for exploring the effects of gravitation on space and time. Let us start with space and consider the relation between the radius of a merry-go-round and its circumference. Normally, that relation is given by the equation $c = 2\pi r$ (c stands now for circumference and r stands for the radius). Imagine someone measuring the circumference of the merry-go-round with a yardstick while it is in motion. The yardstick is a physical object, which means that it undergoes length contraction in the direction of motion compared to a yardstick that does not move in the same direction. Let a second person measure the radius of the merry-go-round with a yardstick as well. Since he moves orthogonally to the motion of the merry-go-round, his yardstick gets perhaps thinner, but not shorter. Thus, since relative to the radius-yardstick, the circumference-yardstick shrinks, it will fit into the circumference more times than the radius-yardstick would. This of course means that the circumference of our merry-go-round is measured to be *bigger* than predicted by the above equation. So, $c > 2\pi r$.

That is odd. The fact that we measure a length (of the circumference) while taking part in an accelerated motion to be different from the value it would have, were it at rest, shows that acceleration influences the shape of the curve. To see this, consider other circles for which the equation $c > 2\pi r$ holds. These are circles drawn on the surface of a saddle, as shown in Figure 11.2 (look at (c) in particular):

In (c), the circumference is larger than it would be in a flat space (a), for a given radius. It seems, then, that acceleration changes the space from flat to saddle-shaped. Moreover, in light of the principle of general relativity, according to which acceleration and gravitation have observationally indistinguishable effects, gravitation can do the same thing. This was Einstein's great new insight: The presence of large enough masses has an effect on the curvature of space (and time).

Clearly, the idea that space and time (or, more precisely, spacetime, because already STR forces us to conceive as space and time as no longer independent; consider how we argued for length contraction from time dilation) can be curved is quite unfamiliar. And yet, given the explanatory and predictive success of both special and general relativity, spacetime curvature seems to be an unavoidable consequence. Therefore, let's look a little more closely at what this means.

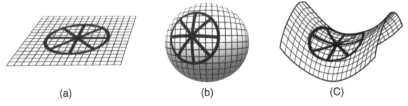

(a)   (b)   (C)

Figure 11.2 Spatial curvature.

The first thing to notice is that spacetime curvature promises to help us circumvent a conflict between special relativity and the account of gravitation embedded in classical mechanics. According to the latter, gravitation is a universal force acting between any two masses (m_1 and m_2) in accordance with the following equation: $G = m_1 m_2 / d^2$ (where d is the distance between the masses). It is important to realize that this law of universal gravitation does not have a temporal parameter, which means that if you change the value of m_1, the force experienced by both masses would change instantaneously. For example, if the sun were to go out of existence in 2 minutes, we would know about it right then, instantaneously, for the gravitational force we experience would change without delay. It might be like flying off a very fast merry-go-round. However, we would *see* the sun disappear only 8 minutes after it happened, because its light takes 8 minutes to get here. Thus, the gravitational information that the sun is gone would have traveled faster than the speed of light, which is of course in conflict with the principle of the maximality of the speed of light in special relativity. Einstein was keenly aware of that, and his theory of general relativity is designed to address this problem. General relativity is, in a word, Einstein's theory of gravitation, intended as a replacement of Newton's theory, which is inconsistent with special relativity.

Why does the earth move along an elliptical orbit around the sun, if the sun is not tethering the earth to it with the gravitational force as a sort of invisible leash? The simple answer is that the earth does not move *around* the sun; it moves in a straight line through space, but a straight line in the vicinity of the sun looks like a closed ellipses due to the curvature of spacetime itself that is induced by the sun. Here is how the gravitational relation between Earth and Moon might be pictured (Figure 11.3 is only an analogy, showing a curved surface embedded in 3D space and spacetime curvature is of course not like that):

However, while this approach avoids the inconsistency between special relativity and classical mechanics, one might object that the notion of curved spacetime is so outlandish that the avoidance of the inconsistency comes at too high a price. We ought not give up our familiar notion of space just so that we can save special relativity. But, fortunately for the defenders of general relativity (if not the advocates of "common sense"), the idea that space can be curved received powerful confirmation in 1919 (Einstein had published his theory of general relativity in 1915). If, as general relativity holds, large masses induce a pronounced curvature in spacetime, we would expect to see indirect evidence for such curvature. For example, we would expect to detect that light, which travels in straight lines, "curves" around a large mass, such as the sun. This means that we should be able to see a star that is located behind the sun, because

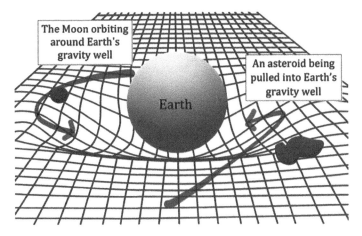

Figure 13.3 Gravitational curvature.

the light curves around the sun, following a straight line in curved spacetime. In other words, if we were to connect our vantage point on Earth through a straight line with some star, and the line crossed the sun, we shouldn't be able to observe the star – unless, of course, space is bent. Thus, if we were able to observe it, this would be strong evidence for the claim that the light emanating from the star "curved" around the sun due to spacetime curvature.

In 1919, Sir Arthur Eddington, a Cambridge cosmologist, led an expedition to the island of Principe off the west coast of Africa, from which he observed, during a complete solar eclipse, just such a star that was located "behind" the sun. Figure 13.4 is a schematic illustration of how this happens:

Spacetime curvature, as an explanation of gravitational effects, seems therefore quite real. Maybe the most exciting consequence of this new theory of gravitation in terms of spacetime curvature is the possibility of traveling backward in time and thus of backward causation. This possibility was first formulated by the Austrian mathematician Kurt Gödel, who proposed a solution to Einstein's field equations that postulated a large-scale structure of the universe in which closed time-like curves are possible.

11.1.4 The Grandfather Paradox

A detailed discussion of Einstein's field equations and Gödel's solution would lead us too far afield, so the following brief remarks must suffice. In a nutshell, Einstein's field equations are mathematical expressions that tell

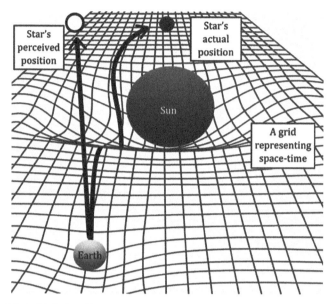

Figure 13.4 Light bending.

us which distribution of mass-energy in the universe is compatible with which spacetime "landscape" (i.e., valleys, mountains, and plains in space-time itself). Gödel found a solution to these equations that looks approximately like what you can see in Figure 13.5:[1]

In this simplification of Gödel's solution, we can think of the universe as divided into an inner and an outer cylinder. Earth is located in the inner cylinder, where the temporal dimension, t, points upward in the picture. In the outer cylinder, t points along the circumference of the circle – it has been flipped (due to the spacetime curvature differences) relative to its direction in the inner cylinder. We start our trip on a spaceship at the higher of the two points on t in the inner cylinder. Going always forward in time, we travel into the outer region and then spiral down the inner cylinder. Finally, we arrive at the lower point of the inner t, thus arriving back home before we started. Globally, we have traveled backward in time, despite the fact that during our entire trip, we locally always traveled forward in time. The difference between the global and the local direction of our trip is grounded in the overall curvature of this universe.

One might worry that problems arising from traveling into the past show that Gödel's solution, which allows for such a journey, must somehow be wrong. This doesn't necessarily mean that they involve mathematical

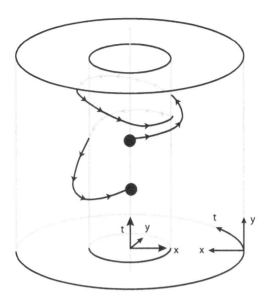

Figure 13.5 Time travel.

mistakes. We can sometimes rule out mathematically acceptable solutions to an equation on independent grounds. For example, suppose you measure the speed of some object as $v = \sqrt{25}$ mph. This equation has two solutions: 5 and -5. However, since no object can have negative speed, you can rule out the second solution on physical grounds, despite the fact that it is a perfectly acceptable mathematical solution. Similarly, it might be the case that there are independent grounds for ruling out Gödel's solution. Indeed, some have argued that certain logical paradoxes that are entailed by time travel into the past provide such independent grounds for ruling out those solutions. The most famous of those paradoxes is the so-called *grandfather paradox*.

Consider Tom, who recently found out that his German grandfather Wilhelm (called Willi by most people) was a key early supporter of the Nazis in the 1930s, providing financial support for them. Tom is ashamed of having such a horrible person as his grandfather (or Opa, as his parents used to call him: "der Willi-Opa"). If only he could have stopped Willi-Opa in his tracks, then maybe the Nazis wouldn't have risen to power and the entire holocaust and WWII, could have been avoided. Tom is angry that he lives now, in 2021, and not back then, when he could have made a difference.

Then, one afternoon, Tom opens his drawer and finds, to his surprise, what looks like a blueprint for some machine. "ZEITMASCHINE" it says in large blue letters across the top. Remembering some German from his childhood, Tom knows what it says: Time machine. Could it be? Tom has studied engineering, so he knows how to read such plans and quickly decides that it's worth a try to put the thing together. After a couple of months (and sizable expense, which he could, however, cover with the money that was for some mysterious reason included with the plans), he managed to build an actual time machine. The moment has come for Tom to act on his disdain for Willi-Opa. He buys a precision rifle, takes some training courses in marksmanship, hops into the time machine, and arrives on a sunny afternoon in 1935 in Munich, Germany.

Tom knows from his parents that Willi-Opa likes to frequent the famous Munich beer gardens. There, he soon finds a person that looks like his grandfather. He follows him home, takes a position across from the house behind some bushes and waits, rifle ready, for Willi-Opa to leave the house again. Suddenly, the door opens, and the grandfather comes out of the house. Tom is ready: He has the training, the weapon, and the will required to kill that bastard. "Finally," Tom thinks, "I can kill him!" But of course, if he succeeded in killing Willi-Opa, his father would have never been born, which means that Tom wouldn't have been born either and he couldn't have traveled back in time to kill his grandfather. Moreover, since Willi-Opa wasn't killed by a sniper on a sunny day in 1935, Tom simply cannot kill him. Past facts cannot be changed. It seems, then, that we arrive at a contradiction: Tom both *can* and *cannot* kill Willi-Opa!

If it were in fact the case that time travel entails contradictions of this sort, then we would have found independent grounds for ruling out Gödel's solutions to Einstein's field equations. We could say that while those solutions are mathematically unassailable, they are illegitimate because they entail contradictions. However, as philosopher David Lewis has argued, the contradiction we identified above is only an apparent contradiction. Here's his argument.

The expression "can" is context relative in the following sense. It is possible that someone can perform a certain action, relative to one set of facts, but also that he cannot perform that action relative to another set of facts. For example, relative to the fact that I am human, have a suitable larynx and the relevant cognitive capacities, I can (in principle) speak Mandarin, while a moth cannot. However, relative to facts about my past, most importantly the fact that I never learned to speak Mandarin, I cannot speak Mandarin. Thus, I can and cannot speak Mandarin. It looks like a contradiction, because the facts relative to which we semantically evaluate the different occurrences of "can" remain implicit. Once they are made explicit, as in the two sentences immediately preceding the apparent contradiction, the appearance of contradictoriness vanishes.

This consideration applies to Tom's situation. Relative to facts about his training, his intention, his functioning rifle, etc., he *can* kill Willi-Opa. But of course, relative to the fact that Willi-Opa wasn't killed that day, Tom *cannot* kill him. Thus, it is no contradiction that he can and cannot kill Willi-Opa. However, one might worry why exactly Tom can't kill his grandfather. Wouldn't pulling the trigger simply change the fact that Willi-Opa wasn't killed that day? The simple answer is that facts can't be changed, at least on Lewis's view. For him, change is a qualitative difference between two distinct temporal parts of the same enduring thing. Facts aren't like that – they have no temporal parts. Therefore, they can't be changed.

From this it follows that if Willi-Opa was not killed in 1935, it is a fact that cannot be changed. Thus, Tom, and indeed nobody, can kill him in 1935. Moreover, if Tom did travel backward in time from 2018 to 1935, and he is in fact waiting behind the bushes with the intent to kill Willi-Opa in 1935, then it has always been the case, from then on, that Tom was there in 1935. In particular, when he, in 2018, hops into his time machine to start his eventually successful trip back to 1935, it was already a fact, at the time that he enters his time machine in 2018, that he was present on that sunny afternoon in Munich in 1935, waiting across from his grandfather's house in a desperate attempt to kill him, the failure of which is also a fact in 2018, when he enters his time machine.

Things become even weirder when we consider how Tom got possession of the plans for the time machine. We said above that he found the plans in his drawer, together with some money to cover the development costs. How did the plans get there? Simple: Tom came back from the future, after he had built the time machine, and deposited the plans (and the money from the future) in his drawer, before he continued his trip back to 1935. But, then, how did the plan arise at all? Where does it come from? The answer is that it didn't "arise" in any familiar way. Finding the plans, building the machine, traveling back in time and depositing the plans in his own drawer are events lying on a closed causal loop. Where does it come from? Who knows? That's how general relativity messes with our everyday metaphysical assumptions about causation. Quantum mechanics will only make things worse.

11.2 Quantum Mechanics and Locality

In this section, we will explore how certain experimental results in quantum mechanics challenge the third condition on causation identified earlier, namely, locality. According to this condition, all causation is local, which means that one cannot cause some event to happen "over *there*" by doing something "over *here*." For example, if I were to clap my hands right now in my home office in Laramie, Wyoming, as I am writing this sentence, my clapping couldn't produce a sound instantly in my friend's office in Philadelphia. But this is the kind of thing that quantum mechanics tells us *can* happen: At least at the level of elementary particles, certain events in my office *can* have instantaneous effects in my friend's office in Philly. Sounds crazy? Well, here are the outlines of the story.

11.2.1 The Measurement Problem

While the exact details can be daunting, the celebrated measurement problem is roughly the following. Being left alone, a quantum system evolves over time in accordance with the so-called Schrödinger equation, which specifies the values of a wave function over time. If, however, we measure the system to determine the value of a certain observable (the equation only gives us probabilistic information – for example, about the location of an electron), it evolves very differently after the measurement. This is sometimes referred to as the *collapse of the wave function*. As philosopher David Albert puts it: "The effect of measuring an observable must necessarily be to *change* the state vector of the measured system, to 'collapse' it, to make it 'jump' from whatever it may have been just prior to the measurement into some eigenvector of the measured observable operator."[2] For example, before you measure it, a particle's so-called spin (the observable in this case) is in a superposition of its two possible values (meaning that neither is actualized), usually called "up" and "down." After measurement, the spin assumes either the value "up" or the value "down" definitively (both are the spin-operator's eigenvectors), which means of course that the state vector representing the system has changed from one in which spin is in a superposition to one in which spin is either "up" or "down."

The question is whether Schrödinger's equation simply *doesn't specify* the electron's definitive spin or whether the electron *doesn't have* a definitive spin ("up" or "down") before we measure it. The latter possibility was horrifying to the likes of Einstein, who subscribed to a *robust realism*. According to this sort of realism, observables, like spin, have definitive values independently of whether or not we measure them. However, if such realism about un-measured spins of electrons is correct, and if Schrödinger's equation only predicts the probabilities with which we measure various positions, quantum mechanics must be incomplete. Consider the following analogy.

Suppose classical mechanics were different and couldn't predict, via Kepler's laws, exactly where Mars is going to be located on June 12, 2045 (recall that location is simply another observable, as was mentioned earlier). Instead, it gives you only a probability distribution over possible locations. But it also says that once you look, you'd see it at exactly one of these possible locations, and definitely not at any of the others. If that was the case, you'd suspect that something was missing, some crucial piece of information which, if only you could find it, would turn the probabilistic prediction into a deterministic one. At any time, Mars must have a definite position, and if classical mechanics can't predict that position, something's missing (what's missing might be called a *hidden variable*). This is exactly the attitude Einstein had toward quantum mechanics, and he devised the following famous thought experiment to show that his attitude was justified.

11.2.2 The Einstein-Podolsky-Rosen Paradox (EPR)

In 1935, Einstein, together with two postdoctoral researchers, Boris Podolsky and Nathan Rosen, published a ground-breaking paper, called "Can Quantum Mechanical Description of Physical Reality Be Considered Complete?" The title was a rhetorical question – Einstein and his associates were sure that this description must be incomplete. The structure of the paper's argument is rather complicated, but its impact was enormous and reverberates until today. As we shall see, it, together with later results that ruled out the existence of hidden variables (knowledge of which would complete quantum mechanics), put serious stress on our ordinary conviction that all causation must be local.

Instead of delving into the intricacies of the original EPR paper, we'll explore a simplified version of the argument that was originally devised by the American physicist David Bohm. It involves a thought experiment concerning the spins of a pair of electrons along different axes. In this experiment, the pair is prepared in a so-called *singlet state*, in which each particle's spin is, of course, in a superposition between "up" and "down," but their respective spins are perfectly anticorrelated: If the spin of particle A is measured to be "up," then the spin of particle B is measured to be "down," and vice versa. Now we imagine that the two particles are separated by being sent off in two different directions until they are as far apart from each other as we may wish. Finally, we measure the spin on particle A, which is predicted by the theory to be "up" with pr = 0.5 or to be "down" with pr = 0.5. Due to the collapse of the wave function (see above), we get a definitive result, let's say, "down." This of course means that B must have spin "up," because the spins remain perfectly anticorrelated. However, we didn't measure B at all, so on account of what does its wave function collapse from a superposition of "up" and "down" to the definitive "up"?

This, in a nutshell, is the problem raised by the EPR paper and Bohm's version. If quantum mechanics were a complete description of the physical universe, then it would be true that before being measured, spins are in a superposition. Since B has not been measured, its spin should therefore remain in a superposition. But due to the perfect anticorrelation between A and B, its spin is definitively "up" after we measure A, which could in principle be light years away from B. Thus, unless we allow for some "spooky action-at-a-distance" (Einstein's own words) between particles that are as far apart from each other as one may wish, quantum mechanics must be incomplete. In other words, there must be some hidden variables determining the precise values of all observables of our system that quantum mechanics does not incorporate. This at least is one way in which to formulate the conclusion of the EPR paper.

This conclusion was, however, shown to be false by American physicist John Bell in 1964. While the details would lead us too far afield, we can say that roughly, Bell's refutation of hidden variables involves an argument showing that if certain observables, such as spin, had definitive values before we measured them, we should see specific correlations in a long run of certain experiments. However, we *do not observe* those correlations. Thus, the suggestion that there might be hidden variables determining a definitive spin before measurement was shown false by testing an observational consequence of assuming that this suggestion is right and showing that this consequence doesn't obtain.

11.2.3 Something Has to Give

Negative results in science are not something to rest on, especially not in this case. Bell's result shows that given certain natural assumptions, quantum mechanics is not incomplete in the sense suggested by the EPR paper. Perhaps then we have to sacrifice our conviction that all causation is local. One way of doing this is by postulating that some sort of signal reaches B instantaneously as soon as our measurement of A's spin reveals that its spin is "down," forcing B's spin to be "up." The problem with this move is that it contradicts a fundamental postulate of the theory of relativity, namely, that no signal can travel faster than light, and certainly not instantaneously. And since, as we have seen, the theory of relativity is well-confirmed, this contradiction is indeed a huge problem.

An alternative to superluminal signaling, considered already by Schrödinger himself in 1935, is to think of systems such as the two electrons that had been prepared in a singlet state before being sent onto their respective journeys as no longer "fully independent" – as, instead, *entangled*. Here's the master himself: "When two systems […] enter into temporary physical interaction […] and when after a time of mutual influence the systems separate again, then they can no longer be described […] by endowing each of them with a [wavefunction] of its own. […] By the interaction the two [wavefunctions] have become entangled."[3]

It is of course one thing to name a potential solution and another one to work out in detail the solution so named, both mathematically and especially conceptually. But here we have to stop our journey into the quantum world, hoping to have indeed whetted your appetite for a deeper engagement with quantum physics through this cliff hanger. It is worth it, although it's not the easiest subject to study. At the end of this chapter, you'll find some suggestions for getting started.

11.2.4 Other Options

There appear to be at least two further ways out of the dilemma introduced by the EPR paper. First, we could reject a influence-based account of causation that results in the contradiction between the need of superluminal influence to account for the EPR correlations and the postulate of the maximality of the speed of light. Not all notions of causation involve a locality condition. According to the *counterfactual analysis*, A causes B if and only if had A not happened, B would not have happened. Nothing in this analysis commits us to the claim that A has to be near B, or that changes in A are communicated to B and bring about changes there. Thus, there is no requirement of locality. This means that, in the end, we may be able to account for the EPR correlations by changing from an "influence" model of causation (one billiard ball influences the motion of another one by bumping into it) to a counterfactual model. These are just speculations, however, and making them more precise by exploring their consequences is far beyond what we can do here.

Perhaps more radical is the option to jettison Einstein's realist attitude toward physical theories. Doing so relieves us of the demand to make sense of what quantum mechanics tells us about the world. Indeed, some physicists have been said to adopt exactly this sort of attitude. For them, quantum mechanics is just a mathematical recipe for predicting the outcomes of various experiments with elementary particles and systems thereof. Instead of worrying about what the world is like according to quantum mechanics, their slogan is: Shut up and calculate! This attitude amounts to a form of *nonrealism* about one of our predictively most successful theories concerning the physical world. To better understand what's at stake when we move from a realist to a nonrealist attitude about our empirical theories (or the other way around, as the case may be), read on.

11.3 Conclusion

We admit it: This chapter was one weird journey through realms of scientific possibility. However, measurements of both temporal durations and spatial distances are relative to the observer, rather than being fixed quantities – and empirical evidence supports this disturbing reality. And so, what exactly precludes us from traveling into the past – if anything? Is causation local or can it stretch across vast regions? Can the idea of entangled systems save local causation? Such scientifically plausible possibilities might make you wonder about reality. If so, you'll be fascinated by the next chapter which raises challenging questions regarding the conventional belief in realism that is held by many scientists. See you there!

Notes

1   Our discussion is indebted to Paul Horwich, Asymmetries in Time. Problems in the Philosophy of Science, MIT Press, 1987,which also provided the model for our illustration.
2   David Albert, Quantum Mechanics and Experience, Harvard University Press, 1992, p.36 (emphasis in the original). The state vector represents a system and its properties, while an eigenvector of an operator applied to the state vector is a new vector pointing in the same direction as the original state vector but can be of a different length.
3   "Discussion of Probability Relations between Separated Systems." Mathematical Proceedings of the Cambridge Philosophical Society, 1935, Vol.31, no. 4, pp. 555–563.

Annotated Bibliography

R. I. G. Hughes, 1992, *The Structure and Interpretation of Quantum Mechanics*.
Harvard University Press, Cambridge, MA. The author discusses the use of Hilbert-space models in quantum mechanics and the difficult interpretive questions they give rise to. The book is a very useful entry into discussions of the philosophical issues in quantum mechanics that are informed by the technical details of the theory.

David Lewis, 1976, "Paradoxes of Time Travel," in *Philosophical Papers Vol. II*: pp. 67–80.
Oxford University Press, 1986. Discusses the grandfather paradox along with the question what change is and why facts cannot be changed – past, present, or future.

Tim Maudlin, 2012/19, *Philosophy of Physics: Space and Time (1); Quantum Mechanics (2)*.
Princeton University Press, Princeton, NJ. These two volumes provide the reader with a thorough introduction to both the special and general theory of relativity and to quantum mechanics. The technical details are held to a minimum, but they are sufficient to appreciate the difficult philosophical problems those theories give rise to. Volume (2) also explains the difference between quantum theory as a recipe for predictions vs. a theory about the fundamental nature of the world.

Leonhard Susskind and Art Friedman, *Special Relativity and Classical Field Theory* (2017) and *Quantum Mechanics* (2014).
Basic Books. A nuts-and-bolts approach to the *Theoretical Minimum* required to understand the mathematics of relativity theory and quantum mechanics. Highly accessible to the motivated reader, it is also accompanied by the Theoretical Minimum Lecture series, which can be found at the You-tube Channel of the Stanford Institute for Theoretical Physics at: https://www.youtube.com/c/stanfordinstitutefortheoreticalphysics.

12

BUT IS ANY OF IT REAL?

12.1 Theories and Truth

In the last chapter, we briefly touched on the question of realism. As we saw there, Einstein's opposition to quantum mechanics derived from what he perceived as its threat to his realistic understanding of the world. Since quantum mechanics doesn't tell us what values certain observables (spin) have before we actually observe them, quantum mechanics must be incomplete, because it is clear – at least so Einstein thought – that observables have their values independently of whether or not we observe (or measure) them. Realism in this sense can be roughly defined as the view that *the world is as it is independently of whether or not we observe it*. Observation is just the means by which we find out how the world really is, had we not been looking. In particular, observation does not bring about any properties in the world that it didn't have before we observed it.

Quantum mechanics threatens this sort of realism because measurement might indeed change the world by bringing about definite values of some observables that they didn't have before we observed them. Suppose John Bell is right (and nobody really doubts this today) that the probabilistic character of quantum mechanics is not just a sign that we have overlooked some important details (an epistemic problem), but rather a reflection of the inherent chanciness in the world (a metaphysical claim). Our observations seem to bring about those definite values. How this happens is deeply mysterious and constitutes the so-called *measurement problem* in quantum

This is Philosophy of Science: An Introduction, First Edition. Franz-Peter Griesmaier and Jeffrey A. Lockwood.

mechanics (see Chapter 11). The slogan "the world is as it is" might not be fully defensible. So, are there any contexts in which realism is plausible?

In quantum mechanics, we saw that some things might lack definitive properties, such as spin, unless those properties are observed. Is this true of other properties we usually take to be uncontroversial, like colors and sounds? And what about the bearers of properties, namely, the objects themselves? For that matter, are all the objects we usually accept as existing independently from our observations, or more generally, independently from our minds, really "out there"? In particular, what about the objects and processes we can't directly observe at all, such as electrons, quarks, force fields, or chemical bonds? What reasons do we have for believing that they exist? This is a question that becomes especially pressing if we have empiricist inclinations.

Once we take quantum mechanics into account, the independent reality of some "things" in the world seems to melt away. So how can we make sure that the whole mind-independent world doesn't melt away as well and morph into just a figment of our imagination, or a construction based on how our cognitive apparatus happens to interact with whatever is "out there?" Fortunately, we don't have to discuss whether there is in fact an independent world; deniers of an independent world are often called *idealists* and believe that all that exists is somehow mind-dependent.

We will take some things for granted in our discussion below – that there is a mind-independent world, that not all forms of observations bring about the things we discover in the world, and that we can discover some features of the world through experimentation and observation. Moreover, we will focus on a specific way of thinking about the world, namely, that approach embodied in our empirical theories. Even these constraints allow for a large number of potential disagreements about what's out there – *really* out there, and not just a product of our imagination, observation, and instrumentation. We will start our discussion by looking at the various theoretical views available within the context of the empirical sciences as answers to the question: What is really out there?

12.2 A Map of the Views

In the empirical sciences, the question of whether certain entities are real or not can usefully be divided into questions about groups and questions about unobservables. Let's suppose for simplicity's sake that typical, observable, mid-sized objects are real, such as a particular dog, a certain tree, and your favorite chair. In other words, we assume for our purposes that observation provides us with sufficiently good reasons for accepting that those mid-sized things are real. Questions about what else is real may arise in two ways. First, we can group such mid-sized objects together into larger classes and ask whether those are real in the sense that they exist as entities over and above the entities grouped together. Examples of such groups are families, forests, and furniture.

Second, the mid-sized objects are often believed to be composed of smaller units, and those units of even smaller ones, all the way down to units that cannot be observed anymore, such as elementary particles. The dog is made up of organs, which are composed of tissues, and then cells, organelles, molecules, atoms, protons/electrons/neutrons, quarks, and strings. Moreover, the behavior of the mid-sized objects is sometimes said to be explicable in terms of unobservable agents – atoms, forces, fields, thoughts, or what have you. For both kinds of unobservable objects, the question of their reality arises.

The *reality of groups* of mid-sized objects gives rise to a debate between *realists* and *conventionalists*. The first camp believes that there are at least some groups that have mind-independent reality; biological species might be one example. The second camp views all groups as simply based on conventions, which license a grouping together of mid-sized objects that is not grounded in nature itself, but may simply reflect the interests of those who are responsible for the grouping – namely humans. An example here might be the group composed of edible plants.

The *reality of unobservables* gives rise to a slightly different debate, where we can designate the two camps as *realists* and *nonrealists*. The former think that certain features of our theories permit us to rationally believe that some unobservables are real, despite the fact that we can't see them, e.g., electrons. The nonrealists can be further divided into those that deny the existence of unobservables (antirealists), those that think we simply cannot know whether or not they exist (empiricists), and finally those who think that the introduction of unobservables in some theories is simply a way of talking that should not be taken literally (fictionalists). In Figure 12.1, we summarize these distinctions (you can think of the three realist positions as affirming the existence of the relevant entities):

In the next section, we will discuss the realism–conventionalism debate about groups, concepts (1) and (2) in our chart. Following that, we will look at clear statements of (3) through (6) and evaluate arguments for and against those different views. As we'll see, many *arguments against* (3) take the form of *attacks on arguments for* (3), suggesting that realism about unobservables is problematic and certainly not the default position. In other words, because the empirical sciences are supposed to be grounded in empirical evidence, postulating unobservables is taken to be suspicious, unless one can provide good reasons for doing so. Think of this as a kind of "guilty until proven innocent" approach by empiricists. Clearly, those reasons will have to be different from ordinary reasons for believing that observable things exist, which can be as simple as our observation of those things.

12.3 Are Groups Real?

Consider an animal species, such as the dog, *Canis familiaris*. In virtue of what do we group together the organisms we collectively call dogs? Are there natural facts about those organisms in virtue of which they form the species "dog"? The idea behind the notion of natural facts is fairly clear when we look at examples of nonnatural facts used to assemble organisms into a group, such as the group of all nonhuman mammals currently in your hometown. There is no reason for supposing that this group is in any way a "natural group." There are no natural facts that connect these organisms. That they are all right now in your hometown seems to be independent from what they are as organisms.

Maybe, we should look for some features of the organisms as possible grounds for grouping them together. One thing that comes to mind is

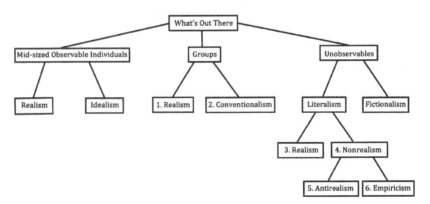

Figure 12.1 Views of ontology.

appearance. Clearly, what organisms look like is not conventional. Cats look more or less like other cats, independently of what anybody thinks. However, a pug, e.g., doesn't look a lot like a Great Dane. To be sure, they have some characteristics in common: four legs, fur, a head, and so on. But those characteristics they share with cats as well. Clearly, if we are to find some features of the organisms that ground natural groupings into different species, those features are not going to be primarily visual features.

12.3.1 The Biological Species Concept

How about behavioral features? Cats behave in cat-like ways: They are often quiet, rather stealthy, and quite independent. Dogs on the other hand are loud, constantly in need of attention, and are protective of their owners and territory. Unfortunately, many other species share these behavioral characteristics (e.g., some children, as well as various birds). However, there might be one kind of behavior that's more characteristic of the individuals making up a species – they interbreed with each other and thereby produce viable offspring. Biologists have taken this observation and proposed the following definition of a species (roughly speaking): Two organisms are members of the same species if they can interbreed and produce nonsterile offspring.

   While clearly grounded in a nonconventional characteristic of organisms, the definition faces two major problems. First, it is inapplicable to species that reproduce asexually, such as some reptiles, many plants and all microbes (the majority of life on Earth!), and those that reproduce both sexually and asexually, such as many aphids. And second, it is also inapplicable to extinct species, such as kinds of dinosaurs. With respect to those, one would have to speak counterfactually: If two members of the species *Tyrannosaurus rex* were to interbreed, they would produce viable offspring. Clearly, empirically minded biologists won't be entirely happy with that sort of move. However, the move might even be required in the case of nonextinct (extant) species. Consider toddlers – they can't interbreed with other humans yet, but they already belong to the same species. Or consider the elderly: Many of them can't interbreed anymore, but are still humans. The same goes for the victims of certain unfortunate accidents or illnesses. Moreover, if Bob never actually visits Japan, he can't interbreed with any of those Japanese who never leave their country – should we therefore assign Bob and those Japanese folks to different species? Obviously not. But in order to avoid this outcome, we again need to introduce counterfactuals: Had Bob visited Japan and met an appropriate mate, he could have interbred with her. The same goes for toddlers, the elderly, and victims of unfortunate circumstances: If things were just a little different (the toddlers a bit older, the elderly a bit younger, and the others a bit more careful or lucky), they all could interbreed with other humans. Counterfactuals abound.

Aside from counterfactuals being unwelcome by empiricists, their usefulness for the cases described above opens the door to another possibility. If we were to create a situation in which a cat could have intercourse with a horse, would they interbreed? We hope that certain phenotypical features, such as size, genitalia, and other anatomical features, are a good proxy for interbreedability. But it's a hope and not clearly grounded in empirical data. Moreover, with artificial insemination, and in particular with genetic modification, all bets might be off. What started as a promising idea is getting pretty murky.

12.3.2 The Phylogenetic Species Concept

An alternative definition of species considers interbreedability as a kind of indirect evidence for what we really believe makes two organisms members of the same species. According to the phylogenetic species concept, individuals belong to a species by virtue of having descended from a common ancestor. So a shared, evolutionary history underwrites the natural kind that we recognize as a species. The challenge becomes how we can infer this unobservable feature (i.e., descent from a common ancestor). This is accomplished by finding one or more unique traits among individuals that we believe represent a particular lineage. The ability to interbreed is not a requisite trait, so the phylogenetic species concept can apply to asexually reproducing organisms.

There are two obvious problems with this definition. First, it is reasonable to doubt the reliability of using contemporary traits as evidence of a common ancestor that may be millions of years in the past (very complex algorithms are used to construct phylogenies – and these methods include questionable assumptions, such as evolution having proceeded in the most parsimonious or simple path possible to get from the past to the present). Second, a perceptive biologist can detect slight differences in virtually any group of organisms, and there's no objective means for deciding how little or how much variation is necessary to infer a common ancestor. Hence, biologists tend to be either "lumpers" (having a high standard for distinctive traits and hence lumping organisms into relatively few putative species) or "splitters" (having a low standard for distinctive traits and hence splitting organisms into lots of presumptive species).

At any rate, the idea that some natural fact should be available as grounds for grouping individuals together in a nonconventional way has strong intuitive appeal. Clearly, there is a difference between the *group of all those organisms located within one mile of you right now*, and *the group of all instances of gold*. The first group is accidental, while the other doesn't seem to be accidental.

12.3.3 *From Biological Species to All of Nature*

Moving from biology to the entirety of the natural world, the importance of how scientists identify real groups of entities becomes all the more vivid. As a first proposal, let's assume that we can characterize real groups as those collecions of entities about which we can make universal generalizations that are not just accidentally true. For example, while the universal generalization "All coins in my pocket are quarters" might be true, it surely is only accidentally so. There is nothing in the nature of my pocket that requires that all the coins in it be quarters. On the other hand, "All gold is soluble in aqua regia" seems to be nonaccidental. Being gold is what accounts for an object's solubility in aqua regia (a mixture of nitric and hydrochloric acids). Putting the pieces together, one might propose that *natural* groupings of objects are those that ground nonaccidental universal generalizations, which are often called *laws of nature*. Thus, we have the following proposal on the table. There are certain groups of things, called natural kinds (e.g., species and gold) the members of which are included because they are all subject to the same natural laws. More precisely, two objects are members of the same natural kind if and only if there is one (or more) law of nature that applies to them in virtue of their membership in that group. For example, two pieces of gold are members of the same natural kind because the law, "All gold dissolves in aqua regia" applies to them in virtue of their membership in the group "gold."

Clearly, to define natural kinds in terms of natural laws will only be useful if we have an independent and good definition of natural laws. As you might expect, the literature on the nature of laws is truly vast, so we cannot possibly hope to fully do it justice here. Instead, we will follow one thread in the discussion that renders the idea of law-based kinds initially quite plausible, but ends up in a skepticism about (the usefulness of) laws.

12.4 Laws of Nature

In Chapter 1, we distinguished between statistical inferences and inductive generalizations. Both are inferences from samples to populations, but there is at least one important difference between them. Consider political polling as a typical example of statistical inferences, and suppose that a particular poll or a set of independently conducted polls concerning a state election shows candidate A ahead by 15 percentage points over her rival B. If the polls are well-designed and timely, they might provide us with a pretty strong reason for believing that A will actually win the election. However, what it will surely *not* provide us with is a reason – *any* reason – for believing that it is a law of nature that candidate A is going to be victorious. Contrast this with the inductive generalization involving the ideal pendulum law. From a number of experiments which provide discrete data points about the relation between the length of the pendulum's string and its period, and through a variety of curve-fitting techniques (recall Chapter 7), we arrive at a regularity which we believe to be a law of nature. Why this difference between the two forms of inference?

One initially promising way in which to explain the difference is in terms of their different logical forms, which we already alluded to. "Candidate A will win the next election in state S" is a statement about a particular future event. "All pendula instantiate the formula $T = 2\pi\sqrt{l\,/\,g}$" is a universal generalization not intended to be restricted in space and time. Thus, it seems that we can *identify* laws of nature with *true universal generalizations* about different kinds of things: pendula, gases, electrons, and what have you.

Unfortunately, things are not this straightforward. There are many true universal generalizations that don't seem to qualify as laws. For example, "All solid spheres of gold are less than one-half mile in diameter" seems true enough, but there seems to be nothing that would make it physically impossible to collect enough gold to produce a sphere larger that one half mile in diameter (however, it might well be economically unfeasible). On the other hand, "All solid spheres of uranium are less than one-half mile in diameter" carries a kind of necessity with it. If we were to attempt to produce a solid uranium sphere with a diameter of one-half mile, the effort would literally blow up into our faces, because such a sphere would exceed by far the critical mass of uranium. In other words, despite both statements being true universal generalizations, only the second one is so because there is a law of nature in the background, while the first one is true because of certain economic and practical contingencies.

Moreover, even putative laws of nature might not be entirely universal, once we think of relevant changes in the environment in which the regularities are realized. Consider once again the ideal pendulum law. It is called "ideal" because it isn't completely true of any real-life pendula, such as your grandfather clock. Instead, it approximates such clocks, because it idealizes away from air resistance, friction, the elasticity of the pendulm string, etc., which are always factors in the behavior of physically real pendula. But now suppose we transport a pendulum, the period of which obeys the law pretty closely here on earth, to a planet where the atmosphere is, at very low temperatures, so thick that it approaches a highly viscous liquid, and where rapid swings in global temperature change the atmospheric viscosity regularly. In such a place, a law that idealizes away from the resistance of the medium would be pretty useless. Rather than making the length of the string the independent variable, one would expect to have the best hope for finding a law, if any, by choosing the ambient temperature (or viscosity) as the determinant of the period. The contributions of the changes in temperature, it can easily be imagined, would simply swamp the contributions of the length of the string to the behavior of the pendulum. In this sense, the ideal pendulum law is not fully universal.

One might insist that considerations of this kind are taken care of by introducing the notion of a *boundary condition* – the context in which a law applies. After all, the pendulum law already needs other boundary conditions to be in place in order to be accurate. As mentioned earlier, the angular displacement of the bob cannot exceed a certain value, especially if the bob is attached to the pivot point by a string instead of by a rigid metal bar of some kind. You can't start the bob directly above the pivot point because the string will just collapse as the weight falls straight downward. Although the observation that the ideal pendulum law is accurate only in situations in which the relevant boundary conditions hold is correct, it raises a further, discomforting question: Why shouldn't we treat the conclusion that candidate A will win the next election in state S as a law that happens to be subject to extremely strict boundary conditions (only the next election in only state S with only the current slate of candidates and only the current political opinions among the electorate counts)?

Faced with this sort of problem, sometimes a modal condition has been proposed as the distinctive feature of law-like generalizations. Consider the super-large uranium sphere once more. It cannot be produced because it is necessary that the uranium undergoes a nuclear reaction as soon as we reach the critical mass. Necessities are tied to counterfactuals. If C necessitates E, then if C were the case, E would be the case as well. For example, if reaching the critical mass necessitates the sample of uranium to undergo a nuclear reaction, then if the uranium sample were to reach critical mass, it would undergo a nuclear reaction. Contrast this with so-called accidental generalization, such as "All the coins in my pocket are dimes." If this generalization were a law, the following counterfactual should be true: "If this quarter were in my pocket, it would be a dime." But clearly, that's just patently false. I often carry quarters in my pocket, and they don't turn miraculously into dimes.

By now, we should have gotten used to the idea that it is very difficult to clarify any of the central concepts we use in the empirical sciences without invoking counterfactuals. Causation, controls, and now laws of nature seem to involve this notion in important ways. However, using counterfactuals in our analysis of laws brings with it an extra problem so severe that quite a few authors have suggested to do away with the notion of natural laws altogether. Here is the problem.

12.4.1 Laws and Counterfactuals

Recall that the truth conditions for counterfactuals require that in all nearby possible worlds in which the antecedent is true, the consequent must be true as well. Clearly, in order to count as a nearby world, its "laws of nature" cannot be allowed to be different from what they are in the actual world. Imagine if we did allow that. Consider the counterfactual, "If this piece of chalk were made out of copper, it would conduct electricity," which seems true. If we allowed far away worlds in evaluating the counterfactual, the statement would turn out false, for there are surely plenty of possible worlds in which copper doesn't conduct electricity, including those in which some demon (a cousin of Maxwell's?) stops electron movement in copper. In fact, if all worlds were fair game for the evaluation of counterfactuals, no counterfactual would ever be true (including this one!), for there are assured to be some possible worlds in which the antecedent holds and the consequent is false (just let the world be populated by the right kind of demons). Thus, when we evaluate counterfactuals, we need to stick to nearby worlds in which our laws of nature hold.

You can see the dilemma we are facing. If we want to distinguish between laws and accidental generalization in terms of counterfactuals, we need to restrict the relevant possible worlds to the nearby ones. However, the nearby worlds are surely those where our laws hold. Thus, if I want to know what a law in our world is, I need to see which counterfactuals are made true by those worlds in which our laws hold. The circularity of this approach is blatant.

12.4.2 Replacing Laws by Regularities

As a possible replacement for laws, some have proposed focusing on regularities that are invariant to a greater or lesser extent across different contexts. The main motivation for this came from the philosophy of biology, which gave rise to early skepticism about laws. In particular, it seems difficult to find anything in biology that both: (i) has empirical content and (ii) is universal in the required sense. In the philosophy of physics, there have also been concerns raised about the alleged universality of laws. American philosopher Nancy Cartwright[1] argues that physical laws are either truly universal, but inapplicable to actual objects and processes, or, to make them applicable, are in need of so many boundary conditions as to make them lose their universality (see our earlier discussion). Thus, both kinds of observations – about the absence of any lawlike empirical claims in biology and about the need of ceteris paribus clauses (i.e., "all other things being equal" clauses) in physics – lead many writers to suggest doing away with the concept of laws of nature. In its place, we should focus on various regularities, some of which are more robust than others, and use them in our explanations of individual events.

What becomes of natural kinds, if we view laws as more-or-less robust regularities? Maybe the concept of a natural kind will have to be reinterpreted as a degreed notion as well, with some more robust kinds (defined by robust regularities) and with some less robust kinds. For example, perhaps complex, sexually reproducing animals make for much stronger natural kinds ("better" species) than do simple, asexually reproducing microbes. But instead of speculating about how a theory of natural kinds might be developed along those lines, we turn to the debate between realists and nonrealists about unobservables.

12.5 Is Everything Real Observable?

12.5.1 Observables vs. Unobservables

To approach the question heading this section, let's start by looking at the concept of unobservables. As the name indicates, *unobservables* are those putatively existing, concrete entities and processes that we cannot observe. Examples come readily to mind: protons, electromagnetic fields, forces of various kinds, dark matter, energy, phyla, other people's mental states, and so forth. Why can't we observe those things? Some are too small, like quarks. Others don't interact, either directly or indirectly, with our sense organs – forces and dark matter come to mind. But why should it matter for the existence of a thing whether or not we can observe it? Surely, other people have mental states, such as beliefs; equally surely, there are protons, chemical bonds, and energy. At the very least, whether or not those things exist seems independent from whether or not we observe them.

It is true: the *existence* of unobservables is independent from our observation. But whether or not we are *justified* in believing that they exist might not be (entirely) independent from observation. But doesn't the fact that we can name something show that it must exist? Why should we have the term "proton" if there weren't any protons? The simple answer is that not everything for which we have names in fact exists. We talk about unicorns and Santa Clause, and they certainly don't exist. Moreover, the history of science is littered with names for things we no longer believe exist: phlogiston (the stuff allegedly responsible for combustion), caloric (the stuff heat's "made of"), the luminiferous ether (the stuff carrying light waves) – and the list goes on. Given this fact, we simply *have* to ask which of the terms in our current sciences pick out unobservable things that in fact exist. What is the rational basis for believing that such things as protons exist? This is the central question in the debate between realists and nonrealists about unobservables.

Before we canvass the various answers to this question, it is important to briefly reflect on what we mean by "observable." Understood literally, we can't observe the inside of the sun, or the far side of the moon. Scaling down, we can't observe genes or molecules, let alone atoms. If we take "observable" too stringently, much of what we believe to exist turns out to be unobservable. In order to avoid this conclusion, we have to say something like "is in principle unobservable." This might still include the interior of the sun, but the far side of the moon becomes observable (via a spaceship), as do the planets in other solar systems (via telescopes) and large molecules (via microscopes). There are problems with this idea, though, the most important of which we encountered in an earlier chapter. This is the calibration problem. If we use instruments to enhance the observational powers of humans, and thus increase the realm of the observable, we need to have some assurance that what we observe through those instruments is really out there and not just a side-effect of how the instrument works. Just consider microscopy for a moment. If you look through any microscope, some of what you see are artifacts of the microscope itself, and it takes a *theory* about how the microscope works to tell you what's real and what's just an artifact of the instrument (e.g., radial distortions in optical microscopy and charging artifacts in electron microscopy) or preparation of the specimen (e.g., bubbles trapped under a cover slip). If observation with the naked eye is theory laden, then observation with instruments is theory laden on steroids!

A precise theory about what counts as in principle (un)observable is very difficult to formulate, and nobody has yet managed to do so. Apart from the calibration problem, there are issues about anthropomorphism: Since humans cannot see ultraviolet, but bees can, is ultraviolet in principle observable or not? Or, if your hairs stand up when a thunderstorm is approaching, does that mean you observe an electromagnetic field? It is unclear how to answer those questions in a principled manner. Thus, we are going to put them aside and consider other vexing problems about unobservables.

12.5.2 How to Name Unobservables

The first of these problems arises as soon as we ask the innocent question of how we manage to name unobservables. Those names are known as *theoretical terms*, and they include the terms "proton," "force field," and the like. The question is this: Since we cannot observe the things we are talking about, how in the world do we introduce a theoretical term for them? For most observable things, the answer to a related question is relatively straightforward. We name an object by somehow making it salient – maybe we point at it – and then utter a name. The naming of a newborn is a good example of this. But for that to be successful, we have to be able to point at the object, which thus needs to be observable. If I were to point at a table and say "electron," it is far from clear that I have succeeded to name anything (apart from perhaps the table). Even if it is true that tables are in part made of electrons, my pointing surely didn't manage to make only electrons salient. If anything, it made the entire table salient. It seems, then, that we can't name unobservables by pointing at them and uttering the chosen name.

12.5.3 Describing Unobservables

Is there any other option? The short answer is yes, but it comes with its own problems for which there isn't a generally accepted solution. To get a feel for this option, consider how a detective might describe the murderer: "Given our evidence, she must be about 5'11″, have shoe size 9, have radial loop patterned fingerprints, and be fairly young." The murderer, then, is the person of which all these descriptions are accurate. Thus, it seems that there is a way to identify an object without directly observing it, as long as we can observe some data that single out that object. Generalizing this idea, one might say that a *name refers to that thing which satisfies a list of (collectively) unique descriptions.*

For example, "Moses," on this view, refers to the person who was found floating down the Nile river in a basket as a baby and who led the Israelites out of Egypt. Similarly, "electron" refers to the particle with the smallest electrical charge. This way, we can introduce a theoretical term into our language without having to point at a particular object. Instead of making the object salient by pointing, we identify it as the thing that satisfies a number of (collectively) unique descriptions. However, there is a problem.

To see the problem, imagine that Moses had a little-known brother, Schmoses. Imagine further that it was in fact not Moses, but actually Schmoses, of whom those unique description are true: Schmoses was found as a baby floating on the Nile and led the Israelites out of Egypt. Now, if our name "Moses" refers to the person of whom those descriptions are true, the name "Moses" picks out Schmoses! In fact, we always talked about Schmoses when we used the name "Moses," because it is Schmoses about whom the descriptions are true.

The same goes for theoretical terms, such as the term "electron." If it refers to the particle with the smallest electrical charge, it actually refers to quarks (assuming they can occur on their own), because quarks have the smallest electrical charge ($1/3$ e⁻). This observation generalizes to all theoretical terms: If their reference is determined by what a theory says is true about it, then, as soon as we change our mind – and the theory – in light of new evidence, our theoretical terms refer to different things! On this view, there would be no stable world of unobservables about which we find out new things through experimentation and reasoning. Rather, whenever we change our theories, we change what we are talking about.

Of course, one might think that such radical changes would be a scandal in the sciences that are supposed to reveal deep truths about the world, including those parts we can't observe but which are sometimes responsible for those things we can observe. In response, some have argued that even though we introduce theoretical terms through descriptions, these are sufficient to fix the thing we talk about, even if we later change our mind about how many, if any at all, of those descriptions are true about the thing. For example, we introduced the term "atom" to refer to the smallest and indivisible building blocks of matter, only to declare much later that we have successfully split the atom! On this view, descriptions can fix the referent of a term without being true of the referent. How this is supposed to work exactly is largely an issue in the philosophy of language, and so we won't pursue it any further here.

12.5.4 Maybe It's All Fictional?

A radical reaction to the problems just canvassed is to "go fictional." Instead of worrying about how to talk correctly about unobservables, we could simply declare that the terms that *ostensibly* refer to unobservable but real entities, *do not refer* to any real entities at all! Rather, terms such as "electron," "force field," and the like, are used to talk "about" fictional entities, not different in nature from Sherlock Holmes or unicorns. We have already encountered something like *fictionalism* in our discussion of analogies. For example, Maxwell's flow analogy is explicitly framed as a fictional story. Electric currents behave *as if* they were some sort of idealized fluid. But electric currents are not any sort of fluid, idealized or otherwise.

According to fictionalists, theoretical terms are just like names for fictional entities. The very question of whether those entities exist or not makes about as much sense as when we evaluate a piece of fiction. A story involving Sherlock Holmes is made neither better nor worse by the fact that he doesn't exist. Similarly, the existence of electrons is quite beside the point when we evaluate theories that use that term. The important question is, instead, whether using the term "electron" has any benefits for our theorizing: Does it help systematize a large number of experimental results? Does it help us understand the phenomena we are interested in? Is it useful for churning out predictions?

While some scientists have found one form or another of fictionalism attractive, others have argued that taking this stance is nothing more than a desperate attempt to avoid the hard questions. To be sure, theoretical terms might have all those functions alluded to above – they result in systematization, help with understanding, and aid in the deduction of testable predictions. However, in contrast to works of fiction, theories make claims about the world. At the very least, what they say about observables, in contrast to unobservables, is clearly either true or false. Fictions don't have such points of contact with reality, but theories do. After all, that's how we test them! This raises the question of realism anew: Given that our most successful theories make largely correct predictions, how can we account for this? At the very least, being a fictionalist about the unobservables and a literalist about the observable looks like a very incoherent position to take. Maybe we should explore the other option a bit further, which we call *literalism*.

12.6 Realism vs. Antirealism

According to a scientific literalist, theories are to be taken as making literal claims about the world, including the claims that involve theoretical terms and are thus ostensibly about unobservables. As such, those theories are either true or false. But which is it? For one kind of literalist, the *realist*, our best theories are at least approximately true. For another kind of literalist, the *antirealist*, even our best theories are most likely false. And a third kind of literalist – the *constructive empiricist* – takes a more cautious stance and claims that we simply cannot know whether those theories are true or false. Now let's look at what evidence the scientific realist claims to have for the truth of our best theories, and how the antirealist might attack this alleged evidence.

12.6.1 Is Predictive Success Mere Coincidence?

The realist starts by asking how we should explain the predictive success of our best theories. Wouldn't it be a strange coincidence if this predictive success could be achieved on the basis of false theories? In other words, it seems that we can best explain the predictive success of our theories by assuming that they are (at least approximately) true.

As an analogy, consider a little black box that always informs you correctly of the temperature, whenever you push the T-button. It stands to reason that the box contains some sort of thermometer. If it didn't have a thermometer, it would be a huge coincidence if it always correctly reported the temperature. However, assuming the existence of a thermometer, the correct reports do not look at all coincidental – they are in fact to be expected. Similarly, if our theories are correct in what they say about unobservable causal processes and agents, the fact that they provide us with successful predictions doesn't look at all coincidental – those underlying causal processes and agents bring about the observable events, which the theory predicts. If the theory is right about the unobservables, that it gives correct predictions is, in fact, to be expected.

The realist relies in this case on an inference to the best explanation, or IBE. She claims that the predictive success of our theories is best explained by assuming that they are (at least approximately) true. From this it follows of course that the entities picked out by the relevant theoretical terms must exist, because if they didn't, what the theory says about them couldn't be true. Thus, realism seems to be the only view that doesn't make the predictive success of our best theories a miracle. This line of reasoning has become known as the *no miracles argument*.

12.6.2 Successful but False Theories – the Pessimistic Induction

The nonrealist objects to the realist's conclusion by pointing out that most, if not all, of our past scientific theories have turned out to be false. For example, classical mechanics postulated gravitational attraction – a force that acts instantaneously over large distances – in order to explain the earth's orbit around the sun. However, in light of the special theory of relativity, there cannot be such a force, as it would convey information faster than the speed of light. Thus, classical mechanics, despite all of its predictive success, turned out to be false. And so it is with most other theories: They turned out to be strictly speaking false, but nonetheless enjoyed predictive success – often of a spectacular sort. Clearly, then, truth is not needed for predictive success, which makes the realist's inference from predictive success to truth highly questionable. This sort of reasoning is often called the *pessimistic induction*. Schematically, it looks like this:

(1) All of our best scientific theories from the past were predictively successful to a greater or lesser degree.
(2) Despite their predictive success, all of our best theories from the past have eventually turned out to be strictly speaking false.
(3) Thus, despite their predictive success, our currently best theories will eventually turn out to be strictly speaking false as well.

Obviously, a theory that turns out false, or is very likely to turn out false, shouldn't be believed. Thus, we shouldn't believe, for example, that there are quarks, because they are postulated by a theory, the so-called *standard model* in particle physics, that will eventually turn out to be false. But if we shouldn't believe what those theories say about the unobservable aspects of nature, one can't adopt a realist interpretation of those theories. Nonrealism is the way to go.

A closer look at this argument might give you pause. It seems that the default position about unobservables is that unless we have a positive reason for believing that they exist and have certain properties, we should not believe that they do. That's why the realist's argument from predictive success is undermined by the pessimistic induction. The dialectic looks like this: The reason we have for believing that certain unobservables exist and have particular properties (such as the belief that there are quarks) is that they are postulated by a predictively successful theory. However, that turns out to be not a very strong reason, because our past theories were predictively successful but still false. The predictive success of false theories is an undercutting defeater of any argument for realism about unobservables that invokes predictive success.

The pessimist's skepticism regarding unobservables also lies in what can be *experienced*, at least allegedly. Nobody in the debate about scientific realism supposes that we can have direct experience of the unobservables postulated by our best theories. Were it possible to directly experience quarks (by seeing them, for example), realism about them might well be the default position. But since this is not possible, we need to support realism about quarks by some argument, of which the argument from predictive success is one – perhaps failed – instance. Thus, in a sense, nonrealism about unobservable is the easier position to defend, precisely because it is the default position and only requires its defender to undermine the arguments for realism.

Recently, a new sort of criticism has emerged against the no miracles argument, involving the base-rate fallacy. As you may recall, this fallacy consists in ignoring base-rate information about the distribution of a property in a population of interest. In our current context, the relevant population consists of theories, and the property of interest is truth. The claim is that we cannot assess the probability that a given theory is true, given its predictive success, unless we know what the base-rate of true theories among all of the theories in our population is. But that means that we need to know on independent grounds how many of our theories are true, which is exactly what's at issue in the debate between realists and antirealists. Moreover, the demarcation problem, briefly discussed in Chapter 3, raises its ugly head again, for we need to know which accounts of the world we should include in our population of theories. Should astrology be included? What about psychoanalysis and creationism? Thus, even if we had some way of assessing the base rate of true theories, we'd still get different probabilities of the truth of predictively successful theories, if we include astrology, psychoanalysis, creationism, and countlessly many others, than if we didn't.

12.6.3 Explaining Success

Nevertheless, nonrealism is challenged by the predictive success of our best theories. Realists propose truth as the explanation for that success, but what do nonrealists have to offer? The constructive empiricist Bas van Fraassen has an intriguing answer. He claims that what explains that our best theories are predictively successful is the simple fact that we *discard those theories that are not*. Imagine a situation in which you have two or more theories designed to account for some range of phenomena, and imagine further that those theories explain the phenomena equally well. However, after testing all of them, it turns out that one is predictively more successful than the rest. In this case, we keep the successful one "on the books," as it were, and discard the others. Since this is what we do in science all the time, it should come as no surprise that the theories we "keep on the books" are all predictively successful.

This explanation is in important ways analogous to some evolutionary explanations. In order to understand why a certain species thrives in a particular environment, we might refer to the fact that it is best equipped to compete for resources in that environment and has thus not (yet) been "discarded" by natural selection, as some of its competitors have been (the thriving species is still "on the books," as it were). However, once we make this analogy explicit, we might start wondering whether van Fraassen's argument doesn't miss the point. The realist is not asking why the best theories are "still on the books," but rather, what it is about those theories that makes them predictively successful. Analogously, while we might be able to explain the success of a species by referring to its competitive advantage during natural selection, there is still another legitimate question we can ask: What is it about this species that confers the competitive advantage? In other words, what features of the organisms making up the species contribute to its success?

In a similar vein, the realist asks about the features of particular theories that account for their predictive success, and she thinks that truth is the relevant feature that can provide us with the explanation we want. While van Fraassen's point is well taken – selecting for successful theories explains why only successful ones are "on the books" – he simply addresses the wrong question. Thus, realism is resurrected as the only view that doesn't render the predictive success of our best theories a complete mystery.

12.6.4 The Argument from Underdetermination

At this point, the nonrealist might produce her most powerful weapon yet: The thesis of the *underdetermination of theories by data*. Recall that for any set of data, there are indefinitely many theories that can account for it – all of them are *empirically adequate* (and thus *empirically equivalent* to each other). If we include the ability to predict new data as a criterion for empirical adequacy, the situation doesn't change radically. For any predictively successful theory, there is a plethora of others that are equally successful. Often, these competitor theories are not compatible with each other, which means that not all of them can be correct. Therefore, since even successful predictions underdetermine which theory is correct, we have no reason to choose one of them and take it as the truth about the world.

Recently, a second line of thought has been added to the underdetermination argument against realism. When we look at the history of science, it becomes apparent that there are many examples of situations in which scientists *failed to even conceive* of relevant alternative theories. For example, in the debate over how to explain planetary motion between Newton (gravitation) and the Cartesians (vortices in the ether), the possibility of space both being merged with time into one framework and also having a mass-energy dependent curvature wasn't even on the table. The relevant concepts weren't available to the scientific community in the seventeenth century. Explaining planetary orbits in terms of spacetime curvature was therefore one of those unconceived alternatives. As Kyle Stanford has shown, the history of science is littered with other examples of unconceived alternatives.

We can use those examples as a further argument against scientific realism. Combining the problem of unconceived alternatives with the argument from pessimistic induction, we get an underdetermination argument that seems extremely strong. Schematically, we can reconstruct it as follows:

1. At any point in the history of science, there existed alternatives to prevailing theories that remained unconceived.
2. Thus, even if it looked as if the data favored decisively one theory over its conceived alternatives, it could not have been decided whether the data favored this theory over the unconceived alternatives as well.
3. There are probably unconceived alternatives to the currently prevailing theories (from 1)
4. Thus, even if it looks as if the data favor a currently prevailing theory over its conceived alternatives, it cannot be decided whether the data favor this theory over the unconceived alternatives as well (from 2 and 3).
5. Thus, one should remain agnostic about the truth of the currently prevailing theories.

In a nutshell, this means that since there have always been unconceived alternatives to prevailing theories, the fact that we may now have only one serious contender as the right theory for some phenomenon doesn't mean that it's the right one – we may not have thought yet about the right one. So, even the drawing of realist conclusions about theories that happen to have no actual, serious competitors is unwise. It's better to remain agnostic.

12.7 Structural Realism

The situation looks dire for the realist. The master argument, combining the problems of underdetermination and unconceived alternatives with the pessimistic induction, seems overwhelmingly strong. Given that false theories can be predictively successful, and given that we often failed to even conceive of theoretical alternatives, what possible grounds are left for the realist to defend her stance?

Perhaps the realist can help her case by dialing back on what we should expect from theories. Instead of expecting that they tell us anything in detail about the nature of the inhabitants of the unobserved world, our best theories perhaps deliver only (approximately) true *structural* information. In the chapter on modeling, we already encountered the idea of structural information. Consider the street map of Vienna. What's important is the correctness of the spatial relations among the streets, which are represented by lines, or bars, and names written on them. The real streets are not lines on paper, nor do they have names painted across their surfaces. Similarly, Bohr's model of the hydrogen atom should not be taken seriously beyond the structural information it contains about the spatial relation between the electron and the nucleus. A view called *structural realism* takes the insight into the essence of models as providing structural information and extends it to all claims about unobservables made by theories: What they say about the nature of the things standing in certain relations should not be taken seriously; the only statements with a claim to correctness are those about the relations themselves – the claims about structures.

There have been a number of studies about structural continuities across often deep theoretical changes in fields such as quantum mechanics, models of the solar system, and others. We will look at the structural continuity between classical mechanics and relativity theory in Chapter 15. Such continuities have been used as evidence against the full force of the pessimistic induction. While it can perhaps be conceded that what past theories say about the inhabitants of structures turned out mostly false, the structures themselves are historically much more stable. If in addition it can be shown that it is the structural information which has been essential for the predictive success of those theories, while claims about properties of the inhabitants of structures are irrelevant to the predictions, predictive success might be a sign of having gotten the structures right (setting aside the base-rate problem). Thus, we should be realists about structures.

There is much lively debate in the literature about structural realism, some of which has led to further interesting distinctions, which we can only indicate here. The main distinction is between ontic and epistemic structural realism: Ontic SR is the view that structure is *all there is*, while epistemic SR holds that structure is *all we can know*. Clearly, ontic structural realism carries a strong and perhaps radical metaphysical commitment. In some versions, it denies the fundamentality of objects and properties and holds that the ontologically fundamental entities are structures. While we ordinarily think that the fundamental furniture of the world consists of objects and their properties, some versions of ontic structural realism hold that objects and properties are grounded in structures. This is perhaps similar to some forms of structuralism in sociology, which claim that what properties participants in a social system have is grounded in the relations among the participants. Epistemic SR is much less radical. It simply stays agnostic about the relative fundamentality of structures, objects, and properties, and holds that from our best, i.e., predictively successful, theories, we can learn structural information about the world of unobservables. Maxwell's flow analogy, discussed in Chapter 9, might be a good example of epistemic SR. He claimed that no physical hypotheses about the hidden nature of electrical phenomena are supported by his analogy, but rather that we can understand the interactions between sources of electricity and sinks in terms of the structure postulated. Which of these variants of SR can be best supported at the expense of the others remains to be decided.

12.8 Realism and Explanation

We noted earlier that antirealism is perhaps the default stance relative to unobservables. On the other hand, realism seems to be the default stance for many of us relative to why we pursue science. We engage in scientific theorizing with an eye toward understanding the world around us, which means that we are interested from the outset to find true, or at least approximately true, theories. In other words, our burning desire to understand what makes the world tick predisposes us toward a realist stance. Our best theories, or so it seems, are our best bet to figure out how the world really works. If we can't be guided by our best theories in the endeavor to understand, why do science at all?

To someone consumed by such a desire to deeply understand the world, the nonrealist's insistence that even false theories can be great instruments for predicting and controlling the realm of mid-sized, observable objects feels like cold comfort. Yes, *instrumentalism* is the right attitude if the only thing we want is prediction and control. In our experience this is the case for many scientists and we are sympathetic to the need for those working in applied fields to address very serious issues facing humanity (e.g., climate change, economic depressions, pandemics, and war). However, this problem-solving framework is a far cry from satisfying the desire for truth and deep understanding.

There is a long tradition in philosophy according to which the world around us is fully intelligible, going all the way back to its beginnings with the Pre-Socratics, who were a group of thinkers that replaced a worldview based on mythological accounts of phenomena embedded in a chaotic world, like the world of Homer's stories, with the idea of the world as a *cosmos* – as something that is ordered and intelligible in such a way that our understanding can fully grasp it. If we assume this to be correct, then a tie-breaker between empirically equivalent theories might perhaps reside in their explanatory power, thus at least alleviating the problem of underdetermination. A theory that provides better explanations than its competitors, whether they are actually or just possibly conceived, is *more likely* to be true of a humanly intelligible cosmos. In other words, let's assume that the world around us is such a cosmos, its structure substantially aligned with our understanding, and then see whether we can make any progress. If we can determine which of a bunch of theories that are all predictively successful can provide better understanding in virtue of delivering better explanations, then, in one grandiose inference to the best explanation, we can infer that this theory is the one most likely true. Explanatory power is the road to realism, predicated on the assumption that the cosmos is by its very nature ready to be understood by us.

Less dramatically, this line of thinking may be summarized as follows. The world around us is, in its entirety and at every scale, intrinsically intelligible for beings like us. Theories that do not provide us with a unified, systematic, and full understanding of the world must be false, because of this intrinsic intelligibility of the world. Thus, the quality of the explanations a theory offers, measured by the degree to which it makes the world intelligible, is a sign of its truth. Of course, if this approach is to have any chance of succeeding, we need to have a pretty good grasp, not just of what makes an explanation good, but also of *what makes for an explanation* in the first place. This is the topic of the next chapter.

12.9 Conclusion

The answer to "what's out there?" is, for scientists, marvelously varied and hotly debated. As we've seen, there is a complex but tractable taxonomy of possibilities – and speaking of taxonomy, species provide a compelling case-study of which and whether scientific groupings are real. Although mid-sized objects moving at slow speeds are entirely valid subjects of study, much of science involves the study of entities and processes that are unobservable. And in sympathy with most scientists, various philosophers have tried to defend a credible account of realism, being unwilling to fully embrace science as a wonderfully useful endeavor but one that ultimately relies on fictional accounts of the world. Extending this line of philosophical inquiry, in the next chapter we turn from the problem of existence to the challenge of explanation.

Note

1   1983. How the Laws of Physics Lie. Oxford University Press.

Annotated bibliography

Anjan Chakravartty, 2017, "Scientific Realism," in *Stanford Encyclopedia of Philosophy*, available at https://plato.stanford.edu/entries/scientific-realism.
    This article provides an in-depth discussion of classical arguments and recent developments in the debate over scientific realism. It is highly accessible.

James Ladyman, 2014, "Structural Realism," in *Stanford Encyclopedia of Philosophy*, available at https://plato.stanford.edu/entries/structural-realism.

Discusses in greater detail the distinction between epistemic and ontic structural realism. It also presents applications of ontic structural realism to questions in spacetime physics and quantum theory.

P. Kyle Stanford, 2006, *Exceeding Our Grasp. Science, History, and the Poblem of Unconceived Alternatives*.

Oxford University Press, Oxford, UK. The author develops his new pessimistic induction against realism based on historical facts about the unconceivability of viable alternatives to certain theories at the time of their proposal. The abstract argument is supported by detailed historical case-studies of the search for the basis of heredity, Galton's stirp theory, and Weismann's theory of the germ-plasm.

Bas C. van Fraassen, 1980, *The Scientific Image*. Oxford: Clarendon Press.

This book constitutes a sustained defense of constructive empiricism, a form of antirealism that takes theories to be either literally true or false but contends that there is no rational basis for accepting specific theories as being either. Instead, we should be content with empirically adequate theories. The author also defends the pragmatic model of explanation which we discuss in the next chapter.

13

EXPLANATION AND UNDERSTANDING

The last chapter ended with the suggestion that we use the understanding delivered by a theory to escape the underdetermination problem. At first glance, this idea smacks of anthropocentrism, or worse, subjectivism. Whether or not a theory leads to understanding surely reflects psychological features of the person considering the theory, or at least features of the human cognitive makeup. We have already seen that our observations are partially influenced by facts about how our cognition works. Wouldn't it stand to reason, then, that our understanding is equally influenced by facts about our cognition? If so, the understanding achieved on the basis of a theory is too closely tied to incidental features of human cognition (best case) or to idiosyncratic psychologies of individuals (worst case) to be of any use in guiding us in our search for true theories. Thus, if understanding is to be afforded any importance, we need to find a model of explanation and its connection to understanding, that at least minimizes the influence of such factors.

Without claiming historical fidelity, it is fruitful to see the various models of explanation proposed during the twentieth century as attempts to rid the notion of explanation of human factors. The goal is to find a *logic* of explanation that is largely independent from human cognitive structures. Human understanding would then simply be a state resulting from grasping objectively existing explanatory relations between that which *stands in need* of explanation and that which *provides* the explanation.

The first model emerged in the 1940s through the work of Carl Gustav Hempel and Paul Oppenheim. It has become known as the *Deductive-Nomological*

This is Philosophy of Science: An Introduction, First Edition. Franz-Peter Griesmaier and Jeffrey A. Lockwood.
© 2022 John Wiley & Sons, Inc. Published 2022 by John Wiley & Sons, Inc.

Model (DN-Model), and also as the *Subsumption Model*. Roughly, explanations are modeled as *deductive inferences* of statements describing that which is to be explained from other statements, among which there must be a statement specifying a *law of nature*. Individual phenomena are explained by subsuming them under general laws, and regularities are explained by subsuming them under more general laws with much broader scope.

Soon, the model came under attack from various directions, eventually giving way to *causal* and *unificationist models*. The causal model emphasizes the role of causation in our explanations, while the unificationist model focuses on the desire to arrive at highly unified theories.

Finally, we will encounter the most pessimistic view of explanation, which is called the *pragmatic model*, but might well be dubbed the *nihilist model*. It emphasizes the influence of contextual factors on what counts as an acceptable explanation in order to establish the view that there are no generally applicable explanatory features. This model appears to get us back to square one, where empirical adequacy is all we can hope for.

13.1 The Deductive-Nomological (DN) Model

One way of understanding the guiding idea behind the DN-Model is that the universe is fully law-governed (nomological) in the sense that no events or phenomena occur that do not instantiate some such laws. Puzzlement over why something occurred results from not seeing the connection between it and the *nomological structure* of the universe. Thus, to explain something is to show how it fits into the nomological structure. This is achieved by identifying the law(s) under which the event or phenomenon in question can be subsumed (hence the name "subsumption model"). Subsuming an event or a phenomenon under a law simply means that it can be *deduced* from a statement of that law together with other statements which specify the conditions that make the law applicable. In this way, the person seeking the explanation is shown how the event or phenomenon fits into the nomological structure of our universe.

In order to make the logic of this model more explicit, we need to introduce a few technical terms. First, as an umbrella term for that which needs to be explained, the term *explanandum* (pl. *explananda*) has been coined. Second, whatever is used to explain a given explanandum is called the *explanans* (pl. *explanantia*). We also have to remind ourselves of the definition of a *deductively sound argument* from Chapter 1: An argument is deductively sound iff it is impossible that the conclusion is false while all the premises are true (validity) *and* all the premises are true (soundness). Finally, since we are going to require that laws play an essential role in an explanation, we need a notion of *relevance* for the premises of an argument. A premise of a deductively valid argument is *deductively relevant* iff the argument would not be valid without the inclusion of that premise. With these terms in hand, we can now provide a fairly rigorous definition of an explanation according to the DN-Model.

> A scientific explanation is a deductively sound argument, the conclusion of which is the explanandum of interest, and whose premises contain at least one deductively relevant law-statement.

This is a mouthful, so perhaps an example is in order. Let's make sure you see that Figure 13.1 exemplifies the above definition. Suppose someone wants to know why a particular piece of ice melted. Here's the canonical DN explanation of the event.

The explanandum is the conclusion of a deductively sound argument, with the premises comprising the explanans. Notice that the explanans consists of two parts: a statement specifying a law of nature and additional statements specifying the so-called initial conditions, which largely determine which laws are applicable to the situation at hand. Notice also that the law is clearly deductively relevant – take it away, and the conclusion (the explanandum) no longer follows from the remaining premises (the initial conditions part of the explanans):

> The water in the glass has a temperature of 32°
> F or above.
> This piece of ice was placed into that glass.
> Thus, this piece of ice melted.

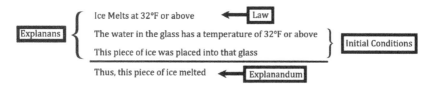

Figure 13.1 The Deductive-nomological model.

Similarly, if we erase the statements specifying the initial conditions, the argument again loses its validity (and thus its soundness):

> Ice melts at 32°F or above.
> Thus, this piece of ice melted.

As one can see, the DN-Model is pretty straightforward, elegant, and goes a long way toward capturing our intuitions about what constitutes a scientific explanation. It makes good on the promise to show how explanations succeed by telling us how some explanandum fits into the nomological structure of the universe: Find an appropriate law and deduce the explanandum from it together with initial conditions. Moreover, the basic structure of the model can be generalized to cover cases in which the relevant law is not strictly deterministic, but statistical instead. The model is called the inductive-statistical model, or IS-Model. Here's what it looks like in an example (Figure 13.2):

The double-line indicates that the argument is no longer deductive, but probabilistic (or inductive). The explanans makes the explanandum probable, but doesn't guarantee its truth. How probable the explanandum is depends on the statistical law. In our example, the higher the probability is that a person recovers quickly after taking vitamin C (e.g., 95/100 cases), the greater the probability of the explanandum. As this probability approaches 1, the IS-Model reduces to the DN-Model, or, in other words, the DN-Model is a special case of the IS-Model, one in which the probability of explanandum/explanans $= 1$.

Hempel and Oppenheim insisted on a high probability requirement for laws used in IS-type explanations. The reason is obvious. If the probability of the law is less than 50%, then it is more likely that the explanandum does not occur than that it does. For example, suppose that hardly any people recover more quickly from a cold after taking vitamin C. Any attempt to explain John's quick recovery by appealing to the fact that he took vitamin C looks

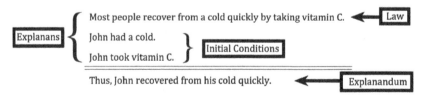

Figure 13.2 The Inductive-statistical model.

like a nonstarter. Taking vitamin C doesn't make him more likely to recover more quickly than not taking it. Thus the high-probability requirement.

Both the IS-Model, and its special case, the DN-Model, appear to be clear winners. However, a closer look reveals that there are many explanations that do not conform to the model, and that there are arguments that should be explanations according to the model, but which intuitively are not. The conditions advanced by the models turns out to be neither necessary nor sufficient for scientific explanantions. Let's consider these shortcomings.

13.1.1 The Flagpole and its Shadow

Shadow reckoning is a dying art. But if we want to know how tall an object is without using a measuring tape, we can apply this ancient method. Simply recall an important proposition from Euclid's *book on geometry*: "If the corresponding angles of two triangles are equal, then the proportions between corresponding sides are equal as well." For example, this proposition can be used to deduce the height of a flagpole from knowledge about the length of the shadow it casts on a sunny day. Figure 13.3 shows how:

Notice the dashed line, indicating the path of the incoming light. It connects the top of the pole with the end point of the shadow. Obviously, the shadow's length depends on the position of the sun. The higher the sun is on the horizon (approaching noon), the shorter the shadow the pole casts will be.

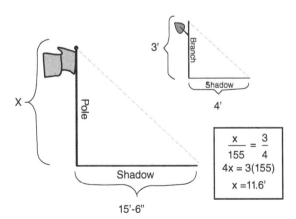

Figure 13.3 Shadow reckoning.

That sunlight travels in straight lines is a law that can be assumed to be true in the context of shadow reckoning. Thus, the length of the shadow depends only on two factors: the height of the pole and the position of the sun, which, together with the laws of optics, determine the length of the shadow.

This suggests that we can explain the length of the shadow in terms of those two factors together with the laws of optics, because such an explanation fits the DN mold exactly. The explanans consists of statements specifying the relevant optical laws and those specifying the initial conditions, viz., the height of the pole and the position of the sun. Thus, what we have here seems to be just another example of a successful explanation in the style of the DN-Model. But remember that we initially talked about shadow reckoning, which is the reverse procedure of how we explain the length of the shadow. In shadow reckoning, we use information about the length of the shadow, Euclid's proposition, the laws of optics, and the position of the sun to deduce the height of the pole. And this is where the troubles start.

Deducing the height of the flagpole looks mathematically exactly like a DN-style explanation. The alleged explanans consists of statements about shadow length, Euclid's proposition, the sun's position, and the laws of optics, from which the statement specifying the pole's height deductively follows. Thus, according to the DN-Model, shadow reckoning not only lets us determine the pole's height, *it also explains it*! And that seems clearly wrong. If somebody where to ask why the pole is 11 5/8 feet high, nobody would offer "because it has a shadow that is 15 1/2 feet long" as the explanation. Shadows of objects don't explain their heights.

The problem illustrated by the example of shadow reckoning has come to be known as the *problem of explanatory asymmetry*. In a nutshell, if one thing explains another thing, that latter thing doesn't explain the former thing. If you explain the broken window by the baseball hitting it, you don't explain the baseball hitting it by the breaking of the window. The relation between explanans and explanandum is asymmetric. Deductive relations, however, are not always asymmetric, as the example of shadow reckoning shows. Put another way, there seems to be a difference between *deducing* values of one parameter from those of other parameters by using certain laws, and *explaining* those former parameter values in terms of other parameter values plus laws. Not all deductions are explanations. But which ones are?

Explanations seem to be *directional*. Thus, if we want to include some deductions among our explanations – and we'll see that we should – they must be deductions respecting the directionality of explanation. We should find a model of explanation which has this sort of directionality already built in. Such a model is in fact available, and it is the *causal model*. Before we explore it, a final critique of the IS model can further motivate causation as an account of explanation.

Remember the high-probability requirement for statistical laws used in legitimate IS-type explanations. This requirement, too, faces a counterexample. There is a certain disease, called *paresis* (a partial paralysis of the face), which afflicts about 30% of those patients who suffer from untreated, latent syphilis. Suppose we want to explain why Joseph developed paresis. It seems clear that identifying his untreated, latent syphilis is a good explanation, despite the fact that the probability of developing paresis under those circumstances is significantly lower than 50%. The high-probability requirement is not met by this IS-style explanation of Joseph's paresis, but it does strike us as a good explanation. Why? Perhaps the right thing to say is that suffering from untreated, latent syphilis is a necessary *causal factor* in the development of paresis. Thus, since nobody without the prior condition develops the disorder, we provide at least a *partial explanation* of Joseph's paresis by pointing out that he suffered from syphilis, even if we can't explain why he developed facial paralysis, in contrast to another syphilitic patient who didn't come down with paresis. At any rate, using Joseph's syphilis in the (partial) explanation of his paresis looks like a case of identifying a causal factor.

13.2 The Causal Model

Causation is directional – if A causes B, B doesn't cause A. Thus, if we require that an acceptable explanation identify the causes responsible for the explanandum, acceptable explanations will automatically exhibit explanatory directedness. Here's how. Suppose we explain the broken window in terms of the baseball that hit it. This amounts to identifying the *salient cause* of the breaking of the window. We use the term "salient cause," because there are other causal background conditions that need to be in place as well. For example, had the window been made out of bullet-proof glass, it wouldn't have broken. But we certainly don't think that the fact that the window is not made from bullet-proof glass is the salient cause of its breaking. Usually, we simply take those background conditions for granted. Thus, the explanation of the broken window in terms of the baseball is a ceteris paribus or *all things being equal* explanation. In this context, it means that we assume that the ordinary causal background conditions obtain, and those conditions include that the window is not made from bullet-proof glass.

We will return later to the question of how to characterize a general "all things being equal" assumption.

According to this causal model, *a scientific explanation consists in the identification of the salient cause(s) of the explanandum.* It was principally Wesley Salmon in 1980 who initially defended the importance of this model, saying that "it puts the cause back into 'because.'" It is fairly easy to see how this model handles explanatory asymmetry. Consider the flagpole and its shadow once more. The height of the flagpole explains the length of the shadow, because all things being equal, it *causes* the length of the shadow. On the other hand, the length of the shadow doesn't explain the height of the flagpole, because it *does not cause* the flagpole's height; most likely, the producer of the pole did. Again, the height of the flagpole is not the only cause of the shadow's length. If the sun is at a different position, there will be a shadow with a different length. But taking these causal background conditions for granted, the pole's height is clearly the salient cause of the shadow's length.

The general directedness of explanation is captured in the causal model as well. Brutus' intention to take over the empire caused him to pierce Caesar's heart with a knife, which in turn caused Caesar to die. This causal chain cannot be reversed. Caesar's death didn't cause his heart to be pierced by the knife. Thus, there is no danger for the model to be exposed to counterexamples that involve reversing deductive relations, in the way in which the DN-Model was. Moreover, the paresis case can be accounted for by using partial causal explanations.

However, the causal model faces its own counterexamples. Not all explanations in the sciences are plausibly reconstructed as causal explanations. In fact, there are many so-called *theoretical explanations* which seem to defy an analysis in terms of causation. In those explanations, a lower level regularity or law, is explained in terms of some higher-level, and broader, regularity or law. For example, we classically explain Kepler's Laws by deriving them (with some corrections) from the laws of classical mechanics together with certain assumptions about boundary conditions. However, it seems abundantly clear that laws, whatever they may be, are not the sort of things capable of entering into causal relations. Kepler's Laws are *not effects* of the workings of Newton's Laws, on any ordinary understanding of the cause–effect relation. Typical effects are the Compton effect in physics (the change in wave-length of X-rays due to scattering from electrons), or the recency effect in psychology (recent items in a list are generally better remembered than earlier ones). In those cases, we think that at least in principle it is possible to identify a mechanism connecting the cause with the effect. It seems odd though to postulate any mechanism mediating between Newton's Laws and Kepler's Laws. It would be a particular kind of confusion – often called a category mistake – to think of Newton's Laws initiating a process that eventually produces Kepler's Laws.

Faced with this problem, the causal theorist might propose to replace this interactionist picture of causation with the counterfactual analysis. According to that analysis, A causes B iff had A not occurred, B would not have occurred either. Applying it to the relation between Newton's Laws (NL) and Kepler's Laws (KL), the relevant counterfactual could perhaps be formulated as follows: If NL weren't true, KL wouldn't be true either. There are at least two problems with this formulation. First, since KL are derived from NL, the truth of NL is only sufficient, but not necessary, for the truth of KL. In other words, KL could be true, even if NL were not true. Second, neither NL nor KL are strictly true; they are both approximations. It is unclear how to apply the standard counterfactual machinery to statements that are only approximately true. In addition to these technical problems, there is a deeper problem for the empiricist: Since the truth conditions for counterfactuals involve other possible worlds, which are empirically inaccessible, counterfactuals are not compatible with a strict empiricist epistemology.

As we have seen in Chapter 10, empiricists replace the counterfactual concept of causation with the conserved quantity theory, which holds that causal interactions are intersections of world lines during which conserved quantities are exchanged. Quite obviously, this theory of causation is of no help whatsoever with the problem at hand, which is to show how the causal model can capture theoretical explanations. NL and KL are not the sort of things that have associated with them world lines, and they certainly cannot exchange any conserved quantities between them. If we adopt the conserved quantity theory as our preferred theory of causation, the problem of theoretical explanations becomes insurmountable for the causal approach. Something else is needed.

13.3 The Unificationist Model

When we look at examples of theoretical explanations, a pattern becomes apparent. Explaining low-level regularities in terms of higher level ones has an added bonus. Typically, the higher level regularities not only explain a single low-level regularity, but a number of others as well. For example, Newton's Laws not only explain Kepler's Laws, but also Galileo's Laws (concerning free fall and projectile motion), the ideal pendulum law (seen as an instance of a dampened harmonic oscillator), Hooke's Law (concerning springs), and others. The fact that higher level laws typically explain a whole range of lower level regularities is the reason why we also characterized such a higher level laws as *broader* in the last section. They are broader in that they subsume a whole bunch of more specific laws under them.

The unificationist model of explanation, developed primarily by American philosophers Michael Friedman and Philip Kitcher in the 1980s and 1990s, puts this sort of *unification* of lower level laws under higher level laws at center stage. Not only does actual scientific practice related to theoretical explanation exhibit unification, there is, in addition, a rationale for unification that becomes visible when we emphasize the relation between explanation and understanding. To begin with, whatever we have to simply accept as a *brute fact* is something we don't understand. The greater the number of brute facts in our world view, the less intelligible the world is rendered, because it is littered with facts – brute facts – that we do not understand. Thus, if we manage to reduce the number of facts in our worldview that we have to accept as brute, the world becomes much more intelligible.

A good example is provided by the rise of classical mechanics in the seventeenth century. Before Newton's work, there were about eight regularities, or laws, on the books, together with additional phenomena, such as the tides, for all of which we lacked any explanation. After Newton, the number of brute (physical) facts was reduced to just four: The three laws of motion (the law of inertia was arguably already among Galileo's Laws) and the law of universal gravitation. Thus, because the number of *brute facts* in our (physical) world view was *decreased*, the degree of understanding was *increased*. The world's intelligibility, on this view, is inversely related to the number of brute facts.

The quest for unification is an important characteristic of the development of science. For example, the nineteenth century saw at least three important examples of unification: Charles Darwin's theory of evolution (1859) unified the plethora of observable species into a tree of life; Dmitri Mendeleev's periodic table (1869) unified (or at least organized) the millions of different substances found on earth into a table of then 59 known and 7 suspected elements (today the table contains 98 naturally occurring elements and 20 synthesized ones, with the search for more underway); and James Clerk Maxwell's theory of electromagnetism (1873) unified the phenomena of electricity, magnetism, and light. Much more recently, during the 1970s, physicists started to pursue a unification of the three fundamental forces (weak, strong, and electromagnetic forces) in the so-called standard model of particle physics, searching for a model sometimes referred to as the *Grand Unified Theory*.

Despite such an impressive history of increasing unification, it is not entirely clear that the unificationist model can deliver on all fronts. First, consider the flagpole counterexample to the DN-Model. We have seen that it is easily handled by the causal model. What about the unificationist model? At first, it doesn't appear that the explanation of the shadow's length in term of the pole's height appeals to any unifying, broader laws. Yes, there are laws involved (geometrical optics), but the same laws are used for deriving the height of the flagpole from the length of the shadow. Thus, if the laws appeared to provide unifying power, the shadow-to-height derivation would be as explanatory as the height-to-shadow derivation. But that's the wrong result. Moreover, the laws do not constitute the salient part of the explanation, which is the pole's height. This raises the question of how many other things can be explained by the pole's height. Maybe just other shadows? In this case, an appeal to unifying power as the difference between the height explanation of the shadow and the shadow explanation of the height becomes quite dubious.

Kitcher tries to alleviate this problem by introducing the notion of *argument patterns* and their unifying power. It would lead us too far afield to detail this notion and how it is put to work. However, the general idea of how attention to argument patterns rules out shadows as explanations of heights is fairly intuitive: Explanations of the heights of artifacts in terms of the shadows they cast will be exceedingly rare (if there are any at all), while explanations of heights in terms of *design specifications* will be commonplace. Moreover, design specifications will explain a great many features of artifacts. Thus, argument patterns that explain features of artifacts in terms of design specifications prove much more unifying than argument patterns involving shadows cast.

There is perhaps a more pressing problem for the unificationist. So many explanations in the sciences are ostensibly causal that the unificationist will need to show that the force of each one of them can be captured by appealing to unifying argument patterns. To do this is surely a tall order. Moreover, if the resulting argument patterns themselves are not appropriately unified, the victory might be in name only. In other words, the program of capturing causal explanations in terms of argument patterns is only fruitful if those argument patterns themselves are perhaps instances of one grand, underlying argument pattern. The hope for finding such a grand pattern seems slim at best. A *grand unifying pattern* looks like a mere pipedream if those philosophers, like Nancy Cartwright and John Dupré, who argue that the world itself is disunified, are right. If there are in fact no grand unifying principles operative in the world, unificationism as a theory of explanation seems decidedly wrong. Since the discussion between the various camps in this debate is far from over, we will have to leave it at that.

13.4 The Pragmatic Model

Perhaps the most pessimistic model emerges from Bas van Fraassen's emphasis on the *context dependence* of acceptable explanations. The relevant contexts include the interests and perhaps abilities of those who look for explanations. The former are often reflected in the so-called *contrast class* for an explanandum, but can also influence the acceptability of explanations as part of the larger background in which the search for an explanation takes place. A couple of examples should make those ideas clearer.

First, requests for explanations typically take the form of why-questions (sometimes also how-questions): Why did the dinosaurs go extinct? Why do the planets in our solar system follow roughly elliptical orbits? Implicit in such questions are contrast classes – why this rather than that? Here's the example adapted from van Fraassen to make the effect of contrast classes on the acceptability of an explanation explicit: Why did the cougar eat the deer? There are three different contrast classes compatible with this question, which we can indicate by using italics and parenthetical information.

1. Why did the *cougar* eat the deer (in contrast to the bear eating the deer)?
2. Why did the cougar *eat* the deer (in contrast to merely killing it)?
3. Why did the cougar eat the *deer* (in contrast to a snake)?

In ordinary conversations, we indicate our explanatory interests by inflection, or stress, or perhaps by explicitly mentioning the parenthetical information. However we do it, the explanatory interest determines which answers count as satisfactory. If I ask "Why did *the cougar* eat the deer?" and you answer "Because it thought the snake would taste oily," your answer isn't appropriate given my explanatory interest.

There are additional context effects. Suppose you are visiting your parents. After dinner, they go for a short walk, while you relax on the porch. Before they leave, they ask you to turn the porch light off when you go inside so as not to attract moths. Pretty soon thereafter you go inside to watch your favorite TV show. But you don't turn off the light. When your parents return, they ask you, somewhat annoyed, "Why is the porch light on?" Imagine you answer, "It's on because the electrical circuit is being closed by the light switch. You see?" They'd probably be even more annoyed.

In light of these considerations, one starts to wonder whether anything can be said in general about what makes an explanation good, acceptable, or appropriate. Clearly, truth is not sufficient. What you said to your parents about the porch light is true enough, but it still fails to count as an acceptable explanation in that context. Given the often unexpected contextual effects on explanatory acceptability, one might just give up on the project of devising a model of scientific explanations that is widely applicable. Sometimes causal explanations are in order, sometimes unifying ones, and who knows, maybe some explanation seekers really desire to be informed about the relevant laws.

One further conclusion that could be drawn is that due to such context effects, the explanatory power of a theory is not a reliable tie-breaker between otherwise empirically equivalent theories. Whether or not a theory has explanatory power depends on the specific why-question being asked, which in turn depends on the intended contrast class (the cougar example) as well as larger background issues (the porch light example). In general, once we take into account these context effects, it is clearly possible that one of a pair of competing theories might be better in explaining one fact while the other one might be better in explaining a different fact. For example, why the cougar, rather than the bear, ate the deer, might be best explained by one of two competing theories about their ecological relationship, while the other one might be better at explaining why the cougar ate the deer instead of the snake. In any case of this sort, appealing to explanatory power as a tie breaker between otherwise empirically equivalent theories is futile. The hope, expressed earlier, to use explanatory power as a guide to truth, seems utterly squashed.

But maybe not all is lost. Suppose we fix the context and then ask anew which of the relevant theories provides the best answer to the specific why-question. In addition, we can ask which model of explanation is the one most applicable to acceptable answers to this now contextualized why-question. That looks like some progress, even if it's not what we'd originally hoped for.

Even if we fix the context, however, this might not be sufficient for fixing the best explanation in the hope to find in it our guide to the true theory. As van Fraassen put it, the best explanation might the best of a whole lot of pretty bad ones. This shows that the status of inference to the best explanation (IBE) is an open question independent of the contextual effects we worried about earlier. Contextual variance crushes the hope of finding a general formula for what counts as an (or the) appropriate explanation. The new worry arises even if we put aside complications arising from contextual variance. It is related to the dubious epistemic status of considerations such as the simplicity of an explanation, along with its conservatism, fruitfulness, or whatever other extra-empirical virtue one might decide to consider. Unless it can be shown that one of these extra-empirical virtues, or at least a good balance of them, has epistemic weight, the pragmatist's worry appears legitimate. We don't claim that assigning epistemic weight to those virtues is impossible (recall our brief discussion of simplicity in the context of the curve-fitting problem). However, it seems that the burden of proof is squarely on the shoulders of those who want to hitch their realist wagon to the fate of IBE.

This leaves us with the question of whether or not fixing the context might at least help us decide which of the two most prominent models of explanation – the causal or the unificationist model – is more promising in a given context. Consider an example, provided by Salmon, involving differences between contexts that do not reflect different explanatory interests, but different kinds of available, relevant background knowledge instead. Suppose you are holding a helium balloon while sitting in an aircraft just before takeoff. The balloon hovers right above your head. Now imagine the airplane taking off, undergoing significant acceleration. In what direction does the balloon move? To many people's surprise, it moves forward, toward the cockpit.

There are two explanations that one can offer for this fact, one with a causal flavor and one with a unificationist bent. The causal explanation is roughly this. As the plane accelerates, the back cabin wall pushes the air particles in the plane forward, producing an air pressure gradient in the cabin, with the pressure slowly diminishing toward the cockpit. This in turn pushes the helium balloon forward as well, with the balloon moving toward less pressure. This kind of explanation is probably acceptable for an audience with an elementary understanding of classical mechanics. Alternatively, one could explain the same fact in terms of Einstein's equivalence principle, according to which how an object behaves under the influence of acceleration is empirically indistinguishable from how it behaves under the influence of a gravitational field. Since the helium balloon rises in the earth's gravitational field, it moves forward in the plane, whose acceleration has the same effect as a gravitational field. This is a unifying explanation, in that it appeals to a very broad principle in which many phenomena are grounded.

The conclusion we can draw from this example is twofold. First, fixing the context (this time in terms of available background knowledge) makes one kind of explanation more appropriate than the other. Second, and more importantly, it looks as if the causal and the unificationist model aren't competing, but rather complementary, models. In one context, causal explanations might be called for, while in other contexts, unifying explanations are the way to go. We have already seen that causal explanations can easily deal with explanatory asymmetries, which pose a problem – maybe not insurmountable – for unificationists. On the other hand, theoretical explanations are the forte of unificationists and are extremely troublesome for causal theorists. For these reasons, Salmon has called for an ecumenical stance about explanations – both are needed to address different contexts.

13.5 What about Realism?

Does all of this mean that the emphasis on the pragmatics of explanation can be made safe for the realist? Remember that we embarked on our journey into the land of scientific explanations hoping to find a guide to true theories through their explanatory power. The idea was that whichever theory from among those that are empirically equivalent has the greatest explanatory power is most likely the correct one. In the background was the assumption that the universe, as cosmos, is ultimately intelligible to beings like ourselves. What the pragmatist suggests is that there is no such thing as *the* explanatory power of a theory. What counts as an acceptable explanation is highly context dependent, and contexts can vary from each other, at least in terms of explanatory interests and relevant background knowledge. Once we fix the context, it might be possible to single out one explanation as the best one (putting worries about IBE to the side), be it a causal or a unifying explanation. However, it would seem that we do not want to claim that the truth about the world is similarly context dependent, which it would be, were it the case that explanatory power is a proxy for truth. If some phenomenon is best explained by theory A in one context, but by theory B in another context, what should we infer about the world, in particular if A and B are incompatible?

We have already encountered a similar problem in connection with learning from models. Recall the ideal gas model as an account of Boyleian behavior. It postulates that there are no interparticle collisions, among other things. This assumption must be denied, however, when we construct a model that can account for facts about diffusion rates. This leads to the question as to whether or not there are interparticle collisions, given that our best explanation of Boyleian behavior denies it. The standard answer is "no," we do not infer the absence of interparticle collisions from the success of the ideal gas model. Rather, we infer that interparticle collisions are not relevant for the production of Boyleian behavior; they are negligible.

It is, unfortunately for the realist, far from clear whether or not a similar strategy is available when dealing with the pragmatist challenge. Could it be the case that the different theories, or models, that are appropriate in different contexts, differ in what they treat as negligible? This would mean that while theories appropriate for different contexts might be incompatible – they "neglect" different aspects of reality in incompatible ways – they can still be approximately accurate, and at least as accurate as ordinary statements about observable objects which also neglect to mention many of the details involving those objects. For the truth of the claim that the book is on the table, facts about its printing date are clearly negligible. For the truth of the ideal gas model in explaining Boyleian behavior, facts about interparticle collisions are similarly negligible.

Whether or not a systematic theory of negligibility (i) could be worked out in satisfying detail and (ii) would ultimately be of help to the realist, is a question we cannot pursue here given the unsettled nature of the issues. However, a recent view in the realist camp that we considered in the last chapter, so-called *structural realism*, might be understood along those lines. In fact, it is perhaps a view according to which almost every detail of our best scientific theories is negligible, except for structural features of the theory. In other words, whether or not the entities postulated by a theory have any or all of the intrinsic properties with which the theory endows them is not important. The only important thing is that they stand in the relations specified by the theory. We will revisit structural realism as an answer to the problem of scientific revolutions and the attendent paradigm shifts in Chapter 15.

13.6 Conclusion

Perhaps the ideal course in science would involve a teacher saying, "Let me explain…" and a student eventually responding, "I understand!" But then along comes the philosopher who wonders just what is meant by explanation and understanding. In this chapter, we considered a variety of models including explanation as deductive inference involving laws of nature, as causal accounts of natural phenomena, as unifications of various regularities, and as providing contextually adequate accounts. Having set the stage for the possibility of a sublime unification of the sciences through over-arching explanations, we turn next to how the varied "layers" or disciplines might – or might not – work together in the pursuit of fundamental theories.

Annotated Bibliography

Carl G. Hempel and Paul Oppenheim, 1948, "Studies in the Logic of Explanation," in *Philosophy of Science* 15(2): 135–175.

Classic and sweeping study that discusses the deductive-nomological model of explanation, teleological (purpose-related) explanations, and problems connected with the notion of a general law, among other things. The paper started a long debate about various counterexamples to its central claims.

Philipp Kitcher and Wesley Salmon, 1989, *Scientific Explanation*. University of Minnesota Press, Minneapolis, MN.

The volume contains Salmon's detailed "Four Decades of Scientific Explanation" (pp. 3–219) and Kitcher's paper "Explanatory Unification and the Causal Structure of the World" (pp. 410–506), which contains the most detailed defense of the unificationist models, introduces the important concept of argument patterns, and presents an attempt to show how causal explanations can be captured within the unificationist framework.

Michael Strevens, 2011, *Depth: An Account of Scientific Explanation*. Harvard University Press, Cambridge, MA.

This work introduces the so-called *kairetic account* (or difference-making account) which analyzes explanation as specifications of causal difference makers. The book is Important, very accessible, and full of interesting examples, but space didn't permit us to cover it in this chapter.

14

FUNDAMENTAL THEORIES AND THE ORGANIZATION OF SCIENCE

While all empirical sciences share many features, such as their reliance on observable evidence, there are also quite a few differences among them. Maybe one of the most striking differences is that the various sciences often describe the same world at various levels of organization. Take, for example, an ordinary cat. A biologist might be interested in certain structures constituting cats, such as their organs or cells. A chemist, in turn, might investigate the chemical reactions involved in the cat's digestive processes. This makes perfect sense if we think of the cat as an organism, the overall behavior and properties of which can be decomposed into the behavior and properties of its parts.

The idea of decomposition of wholes into parts has a long history in the sciences. One of the first attempts to do so can be found in the writings of Democritus, the early Greek atomist. He claimed that everything in the natural world is made of atoms that move around in a void. The power of this idea lies in the way in which it seems to allow for the reduction of the huge variety of macroproperties, such as the colors, tastes, and sounds of things, to a much smaller number of properties of the underlying stuff.

Let's consider color for a moment. Ever since the seventeenth century, writers such as Robert Boyle and John Locke have argued that color isn't a real property of things that they have independently of being observed. More precisely, Locke and Boyle distinguished between two classes of properties: primary and secondary. They thought that only the first class was comprised by real properties of things as they are when unobserved by us. The secondary

This is Philosophy of Science: An Introduction, First Edition. Franz-Peter Griesmaier and Jeffrey A. Lockwood.
© 2022 John Wiley & Sons, Inc. Published 2022 by John Wiley & Sons, Inc.

properties, on this view, are simply the result of the interaction of primary properties with appropriate sense organs. Take away those sense organs and the secondary properties are gone with them. In contemporary parlance, color is what we perceive as the result of being exposed to electromagnetic radiation of a particular wavelength emanating from an object.

On such a view, the color red can be identified with a specific part of the electromagnetic spectrum, and thus we can *reduce* color to radiation. In a similar way, we can reduce the heat of a gas to its the mean molecular kinetic energy, sound to air pressure waves, and smell and taste to chemical structures which interact with our olfactory and gustatory systems. The seventeenth century, which saw the first attempts to work out these ideas with any precision, is the birth of our modern, reductive approach to the world.

And indeed, this idea is still alive and well. In 1992, the eminent physicist Steven Weinberg, who received a Nobel Prize for his work on unifying two of the fundamental forces of nature, wrote:

> "The reason we give the impression that we think that elementary particle physics is more fundamental than other branches of physics is because it is. [...] But by elementary particle physics being more fundamental I do not mean that it is more mathematically profound or that it is more needed for progress in other fields or anything else but only that it is closer to the point of convergence of all our arrows of explanation."
>
> (Weinberg, *Dreams of a Final Theory*; p. 55)

Expressed here is the hope that eventually, all explanations of the natural world can be grounded in one fundamental theory, which is, in Weinberg's view, elementary particle physics. Whether or not his choice of the ultimate ground of all explanations is correct, what is important for us is the claim that there exists a *fundamental theory*. In a way, this claim is simply the final consequence of the project that started in the seventeenth century: To reduce the dizzying variety of objects and properties we encounter in the world to a small number of theories, perhaps even to one truly unifying and fundamental theory (recall unificationism from the last chapter).

In this chapter, we explore the idea of there being a fundamental theory. An important related idea is that of *reducing* one theory to another one. In fact, there are two different concepts covered by the idea of reduction. On the one hand, be a *synchronic relation* between theories at different levels of organization. This is the concept at issue in the quote from Weinberg, when he says that elementary particle physics is the most fundamental one in the sense that theories at higher levels of organization, such as chemistry, can be (allegedly) reduced to elementary particle physics (it is telling that the quote is found in a chapter called "*Two Cheers for Reductionism*"). On the other hand, reduction can also be a *diachronic* (i.e., historical) *relation* between two theories at the same level of organization, such as classical mechanics and special theory of relativity (STR). It is often said that STR *reduces to* classical mechanics in the *limit* of small velocities. Of course, there are additional diachronic relations between theories, such as a more recent theory *modifying* an older one, or even *replacing* it. We will talk about these additional relations in Chapter 15. For our discussion of synchronic relations, we start with the so-called *Layer Cake Model*.

14.1 The Layer Cake Model

We are interested in the relations between theories describing the world at different spatial scales, or levels of organization. Many people (including Weinberg) share the view that theories at the lowest level of organization are the most fundamental ones in the sense that higher-order theories can be reduced to them. Perhaps the first model to make this connection among levels of organization, reduction, and fundamentality, explicit was the *Layer Cake Model*, introduced by Paul Oppenheim and Hilary Putnam in 1957.

The model consists of six levels such that each level contains objects that are constituted by objects from the immediately lower level (except for the lowest level, of course). As Oppenheim and Putnam point out, it is a highly idealized model, but it serves as a good starting point. The levels, paired with the relevant scientific disciplines, are as follows:

Social Groups Sociology: Economics, etc.
Multicellular living things: Psychology, Biology, etc.
Cells: Biology
Molecules: Chemistry
Atoms: Physics
Elementary Particles: Quantum Mechanics

A few things are important to note. First, the pairing indicated in the model is between levels of organization in the world (*ontological* levels, ontology being the study of which entities exist and how they are organized) and levels of scientific disciplines. Just as the ontological levels are stacked in order of breadth (everything is made out of elementary particles, but not everything is made out of cells), so the discipline levels are stacked in order of scope (quantum mechanics, or QM, is applicable to all objects, while cytology only to those that are constituted by cells).

Second, the higher up the levels we move, the more likely it becomes that more than one discipline deals with objects at that level. While currently, we only have QM for elementary particles, certain organisms, as examples of multicellular living things, can be investigated by psychology, biology (including biochemistry, physiology, anatomy, and behavior), evolutionary theory, and perhaps others. This might be due to the fact that higher-level objects have more internal complexity and thus allow for a variety of different research questions to be asked as a function of this complexity.

And third, the model is idealized in that there might be important interactions among the disciplines that cross ontological levels. For example, economic factors might be relevant to explanations of the prevalence, or even onset, of particular diseases in certain populations, whereas the disease, such as cancer or COVID-19, might be expressed at the cellular level.

What makes the model interesting is its relation to reduction. Oppenheim and Putnam proposed that theories from higher-level disciplines can be reduced to theories from the discipline at the immediately lower level. The idea here is that, say, economic theories can be reduced to psychological theories, which in turn can be reduced to neurological theories (a specialized form of cytology, perhaps), and so on all the way down to QM. If this were right, Weinberg's claim that elementary particle physics is the most fundamental theory would be vindicated: All explanatory arrows indeed would converge on QM. The question is whether or not such an optimistic attitude is justified. In order to investigate it, we need to be more explicit about what is required by the sort of reduction the model invokes.

14.2 Classical Reductionism

One of the darling examples of classical reductionists is the relation between thermodynamics (TD) and statistical mechanics (SM). Thermodynamics deals with the macroscopic behavior of gases. It postulates laws, such as the *ideal gas law*, according to which the pressure (P), volume (V), and temperature (T) of a gas are related in the following manner (k is the so-called gas constant): $P = kT / V$. Pressure and volume are familiar mechanical properties of systems. But what is temperature? Is it a nonmechanical property or can it somehow be shown to be a mechanical property "in disguise"? The answer to the latter is affirmative; temperature *is* a disguised mechanical property. More precisely, the temperature in a gas *is identical with* the mean kinetic energy of the molecules making up the gas. The kinetic energy of any particular gas molecule is a mechanical property (roughly, it's a function of how fast the molecule is moving), and the mean energy is just the average of the energies of a large number of those molecules. Thus, through the *theoretical identification* of a thermodynamic property (temperature), with a mechanical property (mean molecular kinetic energy), we achieve an ontological reduction of thermodynamics to statistical mechanics. In short, temperature is nothing "over and above" kinetic energy. So, we don't need to add it to our list of the fundamental properties of the universe. It is not fundamental. It is just a particular form of mechanical energy.

In addition to the ontological reduction, we also get an *explanatory reduction*. The ideal gas law from above can be reduced to laws of statistical mechanics. We don't need to worry about the details here. Suffice it to say that thermodynamic regularities can be derived from the laws of statistical mechanics (on the basis of certain idealizing assumptions). The upshot of this explanatory reduction is that we do not have to add the ideal gas law as a fundamental law to our physical world view. From this perspective, the ideal gas law is simply a specific instantiation of certain mechanical laws. Just as temperature is not a fundamental property of the universe, so is the ideal gas law not a fundamental law.

We are now in a position to formulate two central constraints on classical reduction. If we let T_1 be the reduced theory, and T_2 the reducing theory, the constraints can be expressed in the following schema:

T_1 *reduces* to T_2 iff

1. *Objects* of T_1 are *constituted* by objects of T_2
2. *Laws* of T_1 are *derivable* from laws of T_2

The relation between TD and SM obviously fits this scheme, which can be readily seen by replacing T_1 by TD, and T_2 by SM.

It is crucially important to note that the derivability of higher-level laws (such as those of TD) from lower-level laws (those in SM) require a *coordination of the vocabularies* used in formulating the respective laws. If the lower-level laws do not contain terms relatable to those used in formulating the higher-level laws, no derivation would be possible. For example, "temperature" is simply not a term that is used in SM, which talks about probabilistic relations among mechanical features of systems of particles, and "temperature," as used in TD, is in no obvious way a mechanical term. If anything, it looks like a term for whatever accounts for thermal experiences, such as heat or cold. This is why, in the reduction of TD to SM, we find the theoretical identification of temperature with mean molecular kinetic energy. Identifications of this sort have been called *bridge principles*, as they coordinate the vocabulary of TD with that of SM:T = mean molecular kinetic energy.

Once we adopt this form, we can formulate similar scientific principles:

lightning = electrical discharge
water = H_2O
visible light = the section of the electromagnetic spectrum from about 380 nm to 700 nm
microevolution = changes in allele frequencies over time within a population

One hopes that for every such theoretical identity, there is the possibility of deriving the laws regulating the behavior of the left-hand term from laws regulating the behavior of the right-hand term. Conversely, whenever we want to find such derivations, we first need to find an appropriate bridge principle. If, for all the levels in the Layer Cake Model, it is at least in principle possible to find appropriate bridge principles connecting the lower-level vocabulary with the higher-level vocabulary, the program of unifying science by carrying out a stepwise reduction of higher-level to lower-level theories looks promising. As an example, consider the claim, often encountered in popular discussions of the issue, that the mind is just the brain, or, more precisely, that psychology can be reduced to neuroscience. Can that really be done, even in principle?

14.2.1 Challenges for Neuroscience: Thought

Some multicelluar organisms can engage in activities that are not in any obvious way connected with their cellular make-up. Consider humans and their ability to engage in sophisticated forms of thinking, such as providing a counterfactual analysis of causal relations between events in the world. We know that neurons in the human cortex are intimately involved in such thinking, and we know that neurons are particular kinds of cells. What we don't know is how exactly the activity of such cells results in thoughts. For the classical reductionist, the problem is particularly vexing – what it would even mean for thoughts to be constituted by neurons. Thoughts certainly don't seem to coincide with regions of space filled by pyramidal cells, the most numerous excitatory cells in our brains. And yet, if the classical reductionist is right, facts about thinking must be derivable from facts about neuronal structures.

To get a better grip on the issues involved here, let's think about the problem as the challenge of reducing cognitive psychology (CP) to neuroscience (NS), and assume that cognitive psychology is about the cognitive activities at the organism level, and neuroscience is about those cell structures that seem to underlie such activities. In terms of our classical reduction schema from above, the problem is that of showing that both conditions 1 and 2 of the following biconditional are satisfied:

CP reduces to NS iff

1. Objects of CP are constituted by objects of NS
2. Laws of CP can be derived from laws of NS

For there to be any hope for satisfying condition 1, we would have to reconceptualize the objects of CP in radical ways, which looks like a daunting task. But even setting aside that problem, trying to satisfy condition 2 turns out to be extremely challenging.

First, we need to find laws on the level of CP. Here's one that philosophers of mind have considered:

> *For any agent A, if A desires to have X, and believes that s/he can't have X without having Y, then, all things being equal, s/he desires to have Y.*

A simple instance of this "law" is the following. Suppose Jim desires to have chocolate ice-cream, and that he believes he can't have chocolate ice-cream without also having enough money for buying it in his wallet. So, everything else being equal (e.g., he doesn't already believe he has enough money), he desires to have enough money in his wallet.

So far, so good. Now we get to the challenging part: Find some laws of NS that allows you to derive this CP "law." Clearly, what is needed, as in all cases of reduction, is a way to coordinate CP-terms with NS-terms. The central CP-terms in the "law" are the terms "believes" and "desires." Thus, to have any chance of a successful reduction, we need to find formulas that look like this:

believing = neural events or processes, …?
desiring = neural events or processes, gut feelings, …?

In other words, we need bridge laws, or theoretical identifications, between terms from CP and NS. What could those latter be? One way of approaching this question is through the study of neuronal processes that are correlated with episodes of believing or desiring. The idea is that if we were able to isolate a particular neuronal process that occurs when and only when somebody believes something, we could perhaps infer that this neuronal process is what constitutes believing. There are many questions lurking behind such an approach: Do we expect to find the same type of process across all episodes of believing, regardless of *what* is being believed? Would that process be very different from the process underlying episodes of desiring? How intersubjectively invariant would we expect such processes to be? We won't explore these and other questions here. Instead, we will look at an allegedly simpler case that has inspired many philosophical discussions. It is the case of physical pain.

14.2.2 Challenges for Neuroscience: Pain

Pain is a psychologically complicated thing. We don't like it – usually, unless we believe it is necessary for attaining something we want, perhaps increased fitness by exercising. We try to avoid it – unless we are masochists seeking a new thrill. We scream when we experience it – unless we are Super-Spartans trying to show off our toughness. Neurologically, pain is perhaps even more complicated. The exact processes that underlie experiences of pain are not well understood. Nevertheless, there is a sense that if any type of mental state can be identified with some neuronal process, pain is a prime candidate. It feels bodily and certainly very different from what we experience when we think about the counterfactuals.

During the 1950s and 1960s, a sort of myth about pain's relation to our neurology exercised the imagination of many philosophers. On the basis of observed correlations between the activity of so-called C-fibers and pain, several authors proposed to simply *identify* pain with C-fiber firings, or CFFs. Let's evaluate this proposal:

pain = CFF

Now, we just need to find some "laws" governing pain, translate pain-talk into CFF-talk, and then try to derive the pain "laws" from the laws governing CFFs. This is obviously a huge project, requiring legions of cognitive psychologists and neuroscientists. But in principle, it seems that for pain at least, we can achieve a classical reduction – and if we can do it for pain, who knows? But there is a snag.

There is a little known underground group dedicated to protecting the rights of alien visitors from outer space: ALF, or the *Alien Liberation Front*. You of course know this because you are a member of ALF. Recently, you have been informed by your local commander that some secret government project has been launched to devise methods for gathering information from the aliens about their intentions. Moreover, there is good evidence from informants that the group leading those efforts has succeeded in capturing some aliens after a recent crash in the desert of New Mexico. Your mission: Infiltrate the holding facility and ensure the safety and well-being of the aliens.

You and your fellow activists, being well-trained, gain access to the facility. Horrified, you observe how the aliens, smallish creatures with big eyes, are being tortured with unfathomable brutality by government agents. Your group rushes into the room to demand loudly that the torture be stopped immediately. "How can you inflict such horrible pain onto these small and lovely creatures? Can't you hear their cries; don't you see the anguish in their eyes? Imagine if someone did this to you!" Somewhat annoyed, the person in charge turns to your group: "They are just faking. They aren't in pain. We are just interested in figuring out how much they have learned about how humans react to torture, and given their performance, they seem to know a lot already." You reply angrily, "And what makes you so sure that they are not in pain?!" The scientist sighs dismissively and answers, "Easy. We all know that pain is CFF. We looked, and these aliens simply have no C-fibers. Obviously, if you ain't got no C-fibers, no C-fibers can be firing, and since pain is that firing, you can't be in pain. End of story." But, of course, the philosophical story hardly ends here.

14.3 Functional Concepts

The deep problem with the proposed identification of pain with CFFs should now be obvious: Pain is somehow tied to behavior, including to expressions of experiencing it. To be clear, pain is not the same as those behaviors, and not even the same as being disposed to behave in certain ways (recall the Super-Spartan). But the experience of pain serves a certain function: It alerts the subject to imminent danger to their physical well-being. Pain leads to almost uncontrollable expressions of a certain kind, perhaps to secure aid from others, and it typically initiates avoidance behavior. In an important sense, it doesn't matter to the function of pain how this is carried out, as long as it happens. Perhaps it is true that in humans, the process carrying out the characteristic function of pain is CFF. But in other kinds of creatures, the same function might be physically realized in other ways: Perhaps the aliens are equipped with different internal structures, the excitation of which has the same function that CFF has in us.

Pain looks like a *functional property*. Its essence is the job it performs for the organism as a whole. Its essence is not grounded in the physical properties of the thing that carries out the job. In Chapter 2, we saw other examples of functional properties, such as *being an engine*. The essence of an engine is its job, i.e., to convert energy into motion. That job can be carried out by physically different systems, such as internal combustion engines, steam engines, electrical engines – and even muscles. Similarly, the job of pain, which is perhaps to alert a conscious subject to imminent danger to its bodily well-being and to initiate corrective behavior, can be carried out by physically different systems.

The relation between functional properties, such as pain, and the physical structures of those organisms that can be in pain, is not identity. Instead, the relevant relation is *realizability*. This simply means that pain states, which have a particular function in the life of an organism, are realized in certain ways by the physical states and processes of that organism. Moreover, it is conceptually possible to have the same function realized in physically different systems, and sometimes this possibility is actualized: Engines are realized differently in physically different systems. The possibility of artificial intelligence is predicated on the idea that cognition itself is best understood functionally, and once understood in this way, cognition may turn out to be also realizable in physically different systems, including computers and humans. The technical term for this one-to-many relation is *multiple realizability*.

14.3.1 The Problem with Disjunctive Laws

"Pain" is not the only functional concept. Other important scientific ones include "gene," "insulator," "catalyst," "solvent," "chemical bond," "wing," and countless others. Many of our ordinary concepts are functional too: "chair," "table," "computer," "car," "shelter," etc. The ubiquity of functional concepts makes the Layer Cake Model and its monolithic stratification of the world seem wildly unrealistic. Reductions of higher-level laws involving such concepts to lower-level laws are probably not going to be forthcoming. An influential argument to that effect was formulated by the philosopher Jerry Fodor. Roughly, it goes like this. If higher-level laws (e.g., CP-laws) involve functional concepts, which can be multiply realized, any attempt to derive them from lower-level laws is doomed. Multiple realization means that at the lower-level, a certain function can be carried out by various processes, and it is very implausible that there could be genuine laws that have disjunctions built into them. Such a "law" would have a form like this one:

"If a piston moves upwards or steam is injected or a current is induced, then a magnetic field results or the pressure rises or the spark plug ignites."

Something of this form doesn't look like a law of nature at all. Thus, the reductionist program, which requires that higher-level laws can be derived from lower-level laws, is dead whenever functional concepts figure in the higher-level laws. We need to replace classical reductionism with another account about the relation between different levels of organization. But what are our options?

Two main, perhaps mutually exclusive, options have been proposed in the literature. These options consist in a *kind-relative "reduction"* on the one hand and in the introduction of the notion of *truly emergent properties* on the other. Roughly speaking, a kind-relative "reduction" acknowledges the multiple realizability of many high-level properties but proposes that for each realizer kind (e.g., steam engines and internal combustion engines) we can achieve something close to a reduction by explaining higher-level properties and processes in terms of relations among lower-level properties and the organization of lower-level processes. Let's see how this might work.

14.4 The Functional Model

Functional concepts specify certain tasks carried out by some system. Engines convert energy into motion. Blood transports metabolic materials. Pain alerts the organism to threats to its physical well-being and initiates corrective behavior. Such *macrolevel activities* are invariably carried out in virtue of the activities of some lower-level components of the relevant system. An internal combustion engine converts chemical energy into motion through a complicated mechanism consisting of pistons, fuel pumps, crankshafts, control valves, etc. Oxygen is transported by the blood in virtue of the hemoglobin in the red blood cells binding oxygen, the plasma in which the cells are carried, and the circulatory system. Pain is, as pointed out earlier, even more complicated and not really well understood at this time. But we think that the function it has is achieved by various activities of components of the central nervous system.

In all of these cases, we explain the macrolevel activity in terms of the *organized activities* of lower-level components. It is important to see that the activities of the components have to be *organized* in the right way for the macrobehavior to manifest itself. Those of us who try to fix an engine but are not automobile mechanics learn quickly what happens if the parts are not put together in the right way – the engine won't start anymore. Thus, not just the components matter – the organization of their activities is essential. This insight, first explicitly formulated by Robert Cummins, allows the deployment of a new explanatory strategy which is very different from explaining higher-level facts by reducing them to lower-level facts in the classical sense. Instead of looking for nomological links between different levels, we think of the macrolevel activities as functions of the organized activities of lower-level components and explain them by identifying those lower-level activities and their organization. An example will illustrate this strategy.

Suppose you want to build a chess-machine out of transistors and integrated circuits. Those components can do some interesting things, but individually, they certainly can't play chess. Transistors, however, can be connected into a so-called AND-gate, which computes directly a particular truth function, called conjunction (&). This function (not to be confused with a function in the mathematical sense) takes two truth-values as input and produces one truth-value as output. For example, if both sentences A and B have the value True, then the sentence A & B has the value True as well. However, if either A or B, or both, have the value False, the sentence A & B has the value False too. We can specify the function in a table, with 1 standing for True, and 0 for False.

A B (A & B) in function notation:

1 1 1   &(1, 1) = 1

1 0 0   &(1, 0) = 0

0 1 0   &(0, 1) = 0

0 0 0   &(0, 0) = 0

If we think of 0 and 1 as representing no current (0) or current (1), the AND gate below can be used to directly compute the &-function (Figure 14.1). There will only be current at **Out**, if there is current at both **A** and **B**.

How does that give us a chess-machine? Surely, computing the &-function is not sufficient for playing chess. To make progress, we now need to look at what actually happens when we play chess. The essential part is this: We want to generate a series of moves which get us from the starting position to a

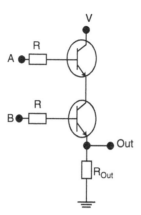

Figure 14.1 AND-gate.

position of checkmate. From this perspective, playing chess looks like computing a function, the CHESS-function. It takes the initial board-positions of the pieces and transforms those stepwise into an array of positions that counts as checkmate. In other words, we are trying to map initial board-positions into a winning board-position. How do we do this?

Obviously, there are many different subfunctions involved: At every move, we need to check the board-positions of the pieces; then we need to generate possible moves, determined by the rules of chess, from which we try to select the move most likely to result in a winning position. Each one of those subfunctions is carried out by further underlying functions. To generate possible moves, we need to consult what the rules allow for the pieces we are considering; to evaluate the possible moves, we need to assess them in terms of some heuristic telling us its strategic value, etc.

The idea behind a full-blown functional analysis is as follows. Consider the macrolevel activity as the stepwise production of a final state of the system from an initial state. Each step can in turn be seen as consisting of other steps, and each one of those as consisting again of further steps, and so on, until we reach a level where a step can be carried out in one fell swoop, without needing intermediary steps anymore. In the chess example, the functional analysis involves analyzing the CHESS-function into the subfunctions of POSITION-CHECK, MOVE-GENERATION, and MOVE-EVALUATION, all of which are analyzed into a sequence of further (sub)subfunctions, until we reach a level of truth-functions that can be directly computed by simple devices, such as the AND-gate from above. Organizing the direct computations of the simplest functions in the right way then produces, at the macrolevel, the behavior we call playing chess. This is, in principle, how playing chess can be explained in terms of the organized activities of components none of which can play chess themselves.

Some such story must be right about human chess players as well. After all, our own neurons cannot play chess by themselves. Thus, it must be the organized activities of those neurons that constitute our ability to play chess, and which may well constitute our ability to engage in all of the other cognitive activities, such as thinking about lunch. We do not explain cognitive activities by deriving cognitive laws from the laws of neuroscience, but instead by showing how organized neurological activities give rise to cognitive activities. This strategy is the core of the functional model.

One important lesson from the functional model is that macrolevel activities and properties are not utterly mysterious, even if they appear to be genuinely novel. Neurons can't play chess, but their organized activities can. Pistons and crankshafts by themselves can't move a car, but their organized activity, together with that of other components, can. We are not saying that all questions about how higher-level activities result from lower-level ones have been answered – far from it. But the functional model seems to provide the outline of a viable research program about how some activities and properties at different levels of organization are related to each other.

14.5 Emergence

As an alternative to the functional model, some have claimed that higher-level properties and processes are sometimes truly emergent in at least two ways: They are *ontologically novel* in that the properties are not possessed by the parts, and they are *epistemically inexplicable*, meaning they cannot be explained even on the basis of complete knowledge of the properties and activities of the parts alone. This perhaps more pessimistic view about the intelligibility of higher-level activities and properties (which we'll collectively call "features") in terms of lower-level features goes under the name of *emergentism*. Frequently, emergentism has been characterized as simply the view that a physical system, once it reaches some appropriate level of complexity, can exhibit features that none of its constituents have. However, this characterization is too vague to show the contrast between emergentism and the functional model of interlevel relations. As we pointed out above, AND gates can't play chess, but when they are combined with other similar circuits and organized in the right way (reaching the appropriate level of complexity), the resulting physical system can play chess.

The real significance of emergentism becomes visible only in the following more stringent formulation: A system-level feature is emergent iff: (i) it is *not possessed* by the system's constituents in isolation **and** (ii) it *cannot be explained* in terms of the system's constituents. An even more radical view results from adding a third condition (iii) that the system's constituents together with their manner of organization *do not necessitate* the system-level features. Condition (iii) entails that if we have two systems that are made from the same (kinds of) constituents organized in the very same way, it is possible that one manifests features that the other one does not. Clearly, condition (iii) defines the strongest sense of emergence, one that leaves the system-level features an utter *mystery*.

14.5.1 The Ontological Condition

Condition (i) can be called an *ontological* condition. It simply says that system-level properties and activities must be *novel* to be emergent. However, some care is needed here. A car has the property of weighing, say, 2 tons, but none of its parts have this property. But we don't want to say that the car's weight is an emergent property. Thus, we need to distinguish between *aggregative* features, such as the weight of a whole system being the sum of the weights of its parts, and emergent properties, which are novel *kinds* of properties. The weight of the car is aggregative, while its capacity to transport a passenger is emergent in that none of the parts have this kind of property by themselves. We should note, however, that some aggregations of even identical parts can give rise to novel properties, as when a sufficient accumulation of uranium constitutes the critical mass necessary to sustain a nuclear chain reaction.

Notice that condition (i) by itself leaves open whether or not novel system-level features can be explained in terms of features of the constituents. We proposed in the last section that at least in principle, many system-level features can be explained in terms of constituent features. And even if we cannot currently explain some features of a system in that way, this does not mean that they are inexplicable. The details might simply be so complex that they currently elude our best explanatory efforts. However, it is perfectly conceivable that there might be some features of highly complex systems that will elude those efforts forever. Humans might simply not be smart enough.

14.5.2 *The Epistemic Condition*

Condition (ii), on the other hand, makes the in-principle inexplicability a further requirement for genuine emergence. It encapsulates the requirement for *epistemic* emergence. Epistemic emergence doesn't just mean that we can't explain the system-level features; it also means that we cannot predict them. From the standpoint of someone who has complete knowledge of the constituents of some system, this unpredictability of novel kinds of features seems baffling. Cases of this sort have been the focus of early theorists of emergence, such as the British philosopher C. D. Broad (1887–1971), who argued against the idea that there is a fundamental science. For him, the central issue was whether the so-called special sciences investigating higher order systems (e.g., psychology or biology) could be distilled into the general sciences (e.g., chemistry and thence to physics). The crucial question became: Is there ultimately only one science with particular cases or are there irreducible differences among the sciences? Broad rejected unification, maintaining that with each level of complexity arose fundamental properties and entirely new regularities. It is *epistemic plus ontological* emergence that gives us this sort of truly interesting emergence.

The ability of a machine to play chess is not emergent in this sense, because we can explain it through functional analysis. This might be true about machines in general. Take two chop sticks. Neither one of them has the causal power of leveraging in isolation. However, if you put one across the top of the other, you end up with a lever. The system consisting of two chop sticks has new causal powers, i.e., it can do what the components (each chop stick) cannot do individually. The leveraging power *emerges* from a system of components that individually do not have this power. We would dare to claim, although we don't have the space to argue for this claim in detail, that the main reason for building machines is to produce, or make emergent, new powers and augmented abilities. Sometimes, as in the pulley, the new powers are simply amplifications of already existing powers. In other cases, such as the lever, we get what seems to be a genuinely different kind of power. However, it doesn't seem to be the case that we cannot explain the emergence of these powers. Any elementary textbook on simple machines (others are the screw, the inclined plane, the wedge, and the wheel plus axle) explains how those machines do what they do. Ontological emergence, to put it somewhat provocatively, happens all around us without being inexplicable. The explanation, though, cannot follow the old Layer-Cake model of reductive explanations. It has to be a functional explanation. Of course, this raises the next question: Are there *any* genuinely emergent features?

C. D. Broad suspected that the features of chemical compounds might be truly emergent. He thought that even from a complete knowledge of the features of hydrogen and oxygen, the liquidity of water couldn't be predicted. To return to an example from psychology, some have argued that *human consciousness* is genuinely emergent in this way. In fact, there are two different ways in which consciousness has been thought to be emergent.

The first way is in line with the combination of ontological and epistemic emergence we just discussed. While the facts concerning consciousness (for example, what dark chololate tastes like to you), on this view, are fully determined by neurological facts,[1] it exhibits novel kinds of features that cannot be explained in terms of those neurological facts. For example, it is difficult to see how the qualities of your experience when you drink a cup of hot tea can be explained in terms of neurological processes. This *explanatory gap* between neurology and psychology is sometimes even used to argue for the much stronger views of substance dualism or property dualism (both of which deny that neurology is the sole determinant of consciousness), but it seems implausible that the mere inexplicability of facts concerning consciousness licenses such an inference (see *This is Philosophy of Mind*, pp. 20ff., for details). However, there might be another problem. If one holds that facts of consciousness are fully determined by neurological facts, but, because of genuine emergence, one cannot explain why this should be the case, then this determining relation between neurology and consciousness itself has to be accepted as a brute fact. It is certainly less than satisfying if we have to include such brute determining relations into our ontology of the universe.

14.5.3 The Mystery Condition

The second way in which human consciousness has been thought to be emergent incorporates condition (iii) from above. In fact, this is a view according to which consciousness is not a function of neurology, because there can be differences in facts concerning consciousness that do not require differences in the underlying neurology. An often-used example is the possible existence of beings of a certain sort, so-called *qualia-zombies*.

The idea is that it is possible that there could be beings that share your neurological make-up, but that lack the qualitative experiences you have when tasting a piece of dark chocolate or receiving a paper cut. They might have the capacity to sense, but not to experience or perceive, what it is like to encounter chocolate or injury. If such human zombies were indeed possible, a broadly physicalist worldview – according to which all the fundamental facts are physical facts, and all other facts are determined (or necessitated) by those fundamental physical facts – would be in trouble. In other words, if it is possible that the same physical facts give rise to different sets of facts about conscious experiences, the physical facts do not fully determine all the facts. Nonphysical facts would have to be included with the physical facts as additional fundamental facts. Obviously, everything turns on the possibility of such qualia-zombies, and a lot of ink has been spilled over this alleged possibility, which we cannot explore in detail. Suffice it to say that we see here again an example of how modal considerations (claims about possibility) play an important role for our scientific worldview. If such zombies are possible, the list of fundamental facts looks very different from the list we get if qualia-zombies are not possible.

14.6 Interdisciplinary Research

In whatever ways different levels are related to each other – emergence, reduction, or functional instantiation – that reality can be described at different scales of organization and degrees of complexity is undeniable. Perhaps this was a contributing factor for the proliferation of specialized scientific disciplines that we observe during the history of science. Aristotle was revered into the late Middle Ages for his natural philosophy, which included everything from physics and meteorology to biology and psychology, to use the current terminology. Especially during the eighteenth and nineteenth centuries, many of our current disciplines branched off and became established as enterprises in their own right. However, while the empirical approaches to the world today are organized into independent disciplines, there is also increasing awareness that specialization has a drawback. Ignorance of results on the part of practitioners of one discipline about results from other disciplines might hinder potentially useful investigations of certain real-world problems. Many basic and applied problems are so complex as to require collaboration of researchers from disparate fields. Consider the ecological and social impacts of climate change. The dynamics and consequences of rising temperatures to ecosystems has a great deal to do with physics, chemistry, atmospheric science, oceanography, biogeography, hydrology, pedology, ecology, genetics, zoology, botany, bio-geography, and microbiology – as well as sociology, anthropology, political science, and economics. How to organize these various disciplines in ways that address such complex problems is the driving question behind recent attempts to understand and promote interdisciplinary research, or IDR.

In the philosophy of science, IDR has not received nearly the amount of attention that matches the hype about its promise as a method of problem solving. Much of the literature to date is concerned with institutional and organizational challenges emerging from IDR. However, there are also a number of methodological challenges with important philosophical ramifications which provide fertile ground for further conceptual research. Imagine, for example, the confusion and conflicts arising from how a physicist and a biologist treat sources of variation in their data, how a hydrologist and a political scientist collect samples, how a cultural anthropologist and a chemist understand "bonding," how a plant ecologist and a sociologist use "community," or even how a zoologist and a microbiologist conceive of "species." We don't want to suggest that those problems are insurmountable, but we wish to highlight their existence which is frequently overlooked. It is at first glance unclear, for example, how neurology could possibly contribute to psychological research about the anxieties that drive certain decision-making processes during a vote about a town's ordinance requiring homeowners to maintain "defensible space" in anticipation of wildfires, if the arguments against the reducibility of psychology to neurology from above withstand scrutiny.

Perhaps the first question to be answered on the road to a comprehensive picture of IDR is whether and how the results of different disciplines can be integrated. According to recent proposals, such an integration can take either the form of a mere combination of results, or a kind of perspectivism. In terms of our earlier example, the combination approach would simply add neurological hypotheses about the realizers of certain mental processes to the psychological hypotheses concerning the influence of anxieties on voting behavior. It is unclear how illuminating such an approach would be. A more sophisticated model of integration can perhaps be derived from perspectivism, a view we briefly introduced in Chapter 1. Recall that this approach provides the means for living with, if not fully resolving or integrating, scientific claims arising from different interests such as assessing whether a forest fire should be allowed to burn. The key lies in clearly explicating the values, concerns and assumptions which circumscribe a perspective regarding a shared problem (e.g., should fire management save jobs, protect homes, conserve rare species, or sustain ecological processes?). As such, an interdisciplinary investigation seemingly must begin with the negotiation of an explicit, unifying perspective – not the single "right" one which doesn't exist, but one that aligns the questions and methods of collaborators.

A fairly new and exciting development within philosophy is the quickly growing literature on conceptual engineering. Roughly speaking, CE is concerned not with the analysis of already existing concepts, but with attempts to create new, or redesign already existing, concepts for reorganizing our experiences of, and interactions with, the world. One of the constraints on CE that has been proposed is the utility of the resulting concepts: New concepts should serve the goals we have in the domain for which those concepts are created. This of course means that there has to be a metaconceptual negotiation about worthwhile goals and their conceptual reverberations within a domain. Perhaps the relations of the new concepts with those that are employed by other disciplines addressing problems within the same domain – think about wide-ranging issues concerning climate change mentioned earlier – could serve as additional constraints on the utility of new concepts. In other words, the perspectivist approach to IDR could perhaps be supplemented by conceptual negotiations among practitioners from different disciplines in order to engineer new concepts that align their perspectives in conspicuous ways. Whether this is achievable or not awaits much further research.

14.7 Conclusion

So, are there any fundamental theories? The question has no easy answer. We have good reason for believing that the Layer-Cake model of interlevel relations is wrong, mostly due to the fact that we find many functional properties at higher levels of organization, which cannot be identified with lower-level properties. This is the problem of multiple realizability. It puts to rest any hope of arriving at a fundamental theory to which all theories about systems at higher levels of organization can be smoothly reduced. At best, we can achieve a piecemeal explanation of system-level features by providing a functional analysis of the system's behavior. Such an analysis, if successful, can even explain ontologically emergent features and thus allow us to understand how novel kinds of, say, causal powers can come into existence. However, if we look at the history of science, we can find many examples of high-level features that eluded our best explanatory efforts for a long time, and some remain inexplicable. Human consciousness might be one such feature, and there are perhaps others, such as the property of being alive. Moreover, what we should count as fundamental facts is not a straightforward empirical question. Whether or not qualia-zombies are possible cannot be empirically decided, but determines our list of fundamental facts. Science, again, turns out to involve philosophical and conceptual considerations in addition to empirical research. In other words, it is not just empirical evidence that determines the course of scientific development; conceptual considerations often play a pivotal role as well.

A particularly striking example of how the development of science can be driven by factors other than empirical evidence emerges from the work of Thomas Kuhn, the eminent historian of science. He argues that during scientific revolutions, there is often disagreement on what counts as relevant scientific evidence. As a result, the outcomes of such revolutions are determined by other factors, including factors that look definitively unscientific. We look at Kuhn's influential account of scientific revolutions and paradigm shifts in the final chapter, which deals with diachronic relations between theories and the question of scientific progress.

Note

1 The technical term for this relation between neurology and consciousness is supervenience. Briefly, a set of facts F supervenes on another set of facts G if and only if there can be no difference in F without a difference in G, but there can be differences in G without differences in F. As an illustration, consider the sine function. There can't be any differences in the y-values without differences in the x-values, but there can be differences in the x-values without differences in the y-values. Thus, the y-values supervene on the x-values.

Annotated Bibliography

Robert Cummins, 1975, "Functional Analysis," in *The Journal of Philosophy*, 72 (20): 741–765.
   The classical source for the functional model of explanation, in which the author proposes to replace the thenprevalent deductive-nomological, or subsumption, model, which cannot account for most explanations in the cognitive sciences.

Sophie Gibb, Robin Findlay Hendry, and Tom Lancaster, 2019, *The Routledge Handbook of Emergence*. Routledge, New York, NY.

The 32 chapters in this book start with foundational discussions of the concept of emergence and then look at emergence in the context of physics, the mind, and the social sciences.

Robert van Gulick and Raphael van Riehl, 2019, "Scientific Reduction," in the *Stanford Encyclopedia of Philosophy*.

available at https://plato.stanford.edu/entries/scientific-reduction. An excellent and extremely detailed overview of the many issues surrounding intertheoretical relations. It considers a number of definitions of the expression "x reduces to y," discusses how reduction plays out in the philosophy of mind, and addresses how it is related to newer debates about grounding in metaphysics.

15

SCIENTIFIC PROGRESS

The scientific world is teeming with activity. Theories are proposed, evaluated, then perhaps accepted, but often rejected. Some of them come to define an entire field of research and are then modified over time with an eye toward improvement. Others are short-lived and quickly replaced by better theories. Of course, even venerable old theories, those that dominated their disciplines for decades, if not centuries, may eventually meet the ultimate fate of being abandoned, perhaps even ridiculed. From the perspective of contemporary chemistry, for example, the phlogiston theory of combustion, reigning for almost a century, looks almost childish. According to this theory, a subtle fluid called *phlogiston* is released from burning substances, which makes it extremely difficult to explain why, after accounting for loss of water, those substances gain weight. In contrast, Newton's classical mechanics seems to fare quite a bit better – we can still use it to calculate the path of fairly slow-moving projectiles (slow compared to the speed of light), although we know that the theory is strictly speaking false and has been replaced by relativity theory. Clearly, then, looking at the history of science, we observe an almost bewildering variety of *diachronic* (sequential or temporal) relations between theories. In this chapter, we will focus on just three of them: theory *modifications*, *replacements*, and wholesale *revolutions*. One of the central questions driving our discussion will be the question of progress: When we look at the history of science, *is there any clear evidence that science progresses?* And what do we even mean by scientific progress? Is it perhaps something similar to what Weinberg expresses in the quote that opened the last chapter, i.e., progress toward the ultimate and fundamental theory about the world?

This is Philosophy of Science: An Introduction, First Edition. Franz-Peter Griesmaier and Jeffrey A. Lockwood.
© 2022 John Wiley & Sons, Inc. Published 2022 by John Wiley & Sons, Inc.

15.1 Science and Technology

It is important to keep distinct the notions of *technological* and of *scientific* progress. Ordinarily, these two notions are often conflated. When asked whether science progresses, people often point to the availability of new gadgets, such as computers and smart phones, which weren't available a mere 100 years ago. Other examples include new medical treatments, improvements in crop production, better cars, bigger airplanes – the list goes on. And since a lot of knowledge in physics, chemistry, biology, and other disciplines is required for the design and implementation of those new technologies, it stands to reason that technological progress is a clear sign of scientific progress. In fact, so the story goes, scientific progress just *is* technological progress.

While it is undeniable that we now have better technologies than our great-grandparents, it would be a mistake to conclude from this that science, which allegedly gives us those technologies, must have progressed. Lots of technologies are independent of scientific research. A celebrated example is the development of the steam engine. About 200 years *before* we had a scientific theory that explains how they work, fully functional steam engines were already available. Clearly, thermodynamics, which explains how such engines work, was not needed for building them. Instead, a lot of tinkering, trial and error, as well as experimentation, led to the engine's development. As some pointedly say, "Technology is not applied science; rather, science is technology explained." In other words, it is often the case that we are able to make technological progress in the absence of relevant and deep scientific understanding. Science, more often than not, has to catch up with our technological progress.

Another way to put the point about the independence of technology from science is this. In many technological endeavors, such as the successful manned moon landing during the Apollo program in 1969, scientific theories that are known to be false can be employed. Given that the speed of even a rocket is very slow compared to the speed of light, we don't need the complicated machinery of the special theory of relativity for computing its trajectory. Classical mechanics is usually sufficient. In this case, from the fact that we could land on the moon using calculations based on classical mechanics, it doesn't follow that classical mechanics is correct. Outdated theories are often applicable in special domains, if the events in those domains can be accounted for as limiting cases of the correct theory. We conclude that technological progress is not automatically evidence for scientific progress, since the former can occur without, and independently of, the latter. Perhaps we can make some headway concerning the question of scientific progress by inquiring into the goals of science. After all, if we have a better idea of what the goals of science might be, we can perhaps also determine whether science gets closer to achieving them. And that would be progress.

15.2 Goals of Science

The notion of scientific *progress*, compared to mere change, embodies a normative assessment. Whether a process is progressive or not depends in general on whether or not it tends to promote some goal(s). This raises the question what the goals of science are, in light of which we can assess whether science progresses, i.e., whether, on the whole, it brings us closer to those goals, or not. In the literature on progress, a wide variety of scientific goals have been proposed: the accumulation of truths about the empirical world; increased explanatory and predictive power; increases in problem-solving power; to answer scientifically significant questions that are unknown today; etc.

It is useful to arrange the goals into four groups. First, science might strive to accumulate more truths, or at least get closer to the truth. We call this the *veritistic goal* (*veritas* being Latin for truth). Second, science might try to give us more knowledge about the world, which can be called the *epistemic goal*. Third, science might aim at providing us with more understanding of the world through better explanations, something that has been called the *noetic goal* (derived from the Greek word for mind, *nous*). And finally, science might intend to increase our problem-solving ability. For want of a better term, we call it the *ability goal*. A few preliminary remarks are in order.

The difference between the veritistic and the epistemic goal lies in the requirement for justification. Knowledge, as you recall from Chapter 1, requires both truth and justification. Thus, if we have only the veritistic goal, it could be achieved serendipitously, i.e., by mere luck. Not so for knowledge; it has to consist of truths that are justified. Justification, in empirical science, is closely tied to its use of evidence. On some views, evidence can confirm a theory, on others, it can, through a failure of falsification, only corroborate a theory. Thus, the epistemic goal is more demanding than the veritistic goal.

On the other hand, the noetic goal seems to be a bit less demanding than the epistemic one. Understanding, although we don't have a precise account of it, seems to be possible in the absence of truth, as we suggested in our discussion of modeling in Chapter 9. Maxwell, for example, claimed that his flow analogy allows us to understand the regularities among certain electric phenomena found by Faraday, but he was adamant that it was merely an analogy and not the truth about electrical currents.

The ability goal is quite different from the rest. Solving scientific problems often demands answers to specific questions that make sense only within a particular framework. For example, working within phlogiston-based chemistry, it might be a scientifically significant question how phlogiston interacts with other substances. After the Chemical Revolution, during which phlogiston was replaced by oxygen as the principle of combustion, this question became meaningless. Thomas Kuhn's cyclical model of scientific development, as we'll see below, suggests that since what counts as successful problem-solving changes during scientific revolutions, in part because old problems become meaningless in the new framework, we should not think of the ability to solve new problems as progress along the lines suggested by the other three kinds of goals.

The progress suggested by those three goals is progress toward more truth, understanding, or knowledge. Starting with Bacon and Descartes in the seventeenth century, that science progresses in this linear way was taken for granted until the 1960s, when Kuhn emphasized the discontinuities in the development of science. Of course, historians of science had long been aware of those discontinuities. However, they were simply seen as progressive leaps, rather than as challenges to optimism about linear scientific progress.

To see why it is prima facie plausible to believe in such linear progress, consider the development of the Special Theory of Relativity (STR). Instead of postulating a troublesome discontinuity between classical mechanics and STR, one might consider the development through the lens of *diachronic reductions*. Not only can scientists sometimes reduce a higher-level theory to a lower-level theory by deriving the laws of the former from those of the latter (using appropriate bridge principles), but they often talk about reducing a *newer* theory to an *older* theory in the *limit* of certain parameters.

15.3 Reduction in the Limit

When scientists say that a newer theory *reduces* to an older theory in some limit, they could with equal justification say that the old theory turns out to be a *special case* of the new one. The following example illustrates nicely this form of reduction. In classical mechanics (CM), momentum is defined as the product of mass and velocity. In a formula, where p stands for momentum, m for mass, and v for velocity, the definition looks like this:

$$p = mv \rightarrow \qquad\qquad (15.1)$$

In the special theory of relativity (SR), the momentum of a particle is defined as follows (m should read m_0 for rest mass, but that's not important for the present point):

$$p = \frac{mv}{\sqrt{1 - \left(\dfrac{v}{c}\right)^2}} \rightarrow \qquad\qquad (15.2)$$

Now imagine that the value of v is really, really small compared to the value of c, which is the speed of light. In that case, $\left(\dfrac{v}{c}\right)$ goes toward zero, as does its square. Thus, the denominator goes toward 1, which means that the relativistic momentum looks more and more like the classical momentum from equation (1). It is in this sense that SR reduces to CM in the limit of vanishingly small velocities (compared to c). Among other things, this fact explains why CM looks like the correct theory for ordinary objects, which never attain anything close to the speed of light. In other words, the physics of slowly moving, ordinary objects is pretty close to being correctly described by CM. This makes it clear in what sense CM is a *special case* of SR.

The special case formulation of this type of reduction allows us to see why the development of science sometimes looks progressive. If a later theory subsumes under it an older theory as a special case, then it looks as if the older theory was "on to something" but only partially. To be sure, its practitioners might not have been aware of the fact that their theory was correct about a special case only. But, as the optimistic story goes, new theoretical developments don't render the old theory false. A special (limit) case of a correct theory is still correct. Therefore, if it could be argued that the majority of new theories typically reduce to their predecessors in the limit of some key parameter (such as the parameter v in the above example), the history of science would recount the development of increasingly general theories from limited, but essentially correct, beginnings.

15.4 How Theories Are Born

In general, however, things are more complicated than the picture of linear progress from limited, but essentially correct, beginnings. As the emergence of new theories shows, while some examples fit this mold, there are many important cases of scientific breakthroughs that don't fit. The contrast is best explored by looking at some relevant examples.

15.4.1 Theory Modification

Perhaps the simplest examples of theory modifications are changes in models. Recall from the chapter on modeling that we often start with very simple models and make them more complicated only if the data require it. Initially assuming a linear relationship between dependent and independent variables, we might construct a model that works approximately in a limited number of cases. If it turns out that further cases do not obey the linearity assumption, we choose a higher-order curve that allows us to optimize goodness-of-fit with the data. In economics, for example, we might start by assuming that there is a direct linear relationship between labor and productivity, e.g., in a factory: The more workers are allowed to use the machines, the more gadgets will be produced. However, as is well known, this linear relationship doesn't hold above a certain threshold. Adding more workers while keeping the number of machines constant eventually has the effect that the workers must wait to use the machines, so that while the overall output is increased, it is not increased in direct proportion to the labor added. This is known as the "law of diminishing returns." Taking this law into account will complicate the model of productivity from a linear to a higher-order curve, which is a simple case of straightforward model modification.

Similar examples can be found in the history of science. Ancient astronomical models shared a number of basic modeling assumptions: (i) the earth is at the center of the universe, (ii) the universe is relatively small, and (iii) the heavenly bodies orbit the earth in concentric circles. However, over time it was discovered that such a simple model didn't fit all the data. For example, the observed retrograde motion of Mars, during which it appears to go backward, didn't fit assumption (iii). Thus, a modification to the basic model was introduced in the form of epicycles: Mars was postulated to circle a point which itself circled the earth, thus explaining, at least qualitatively, the observed motion of Mars. The full Ptolemaic model (around 200 C.E.) incorporated a number of further modifications, which we need not discuss.

Both of the modeling activities we just outlined could fuel an optimistic belief in scientific progress. We start with very simple modeling assumptions and modify them in response to new data. On this view, new models are just more sophisticated variations on a theme. They are a better fit to the data, but the fundamental modeling assumptions remain intact. If all of science had developed like this, scientific progress, through which our theories come ever more closely to the truth without having to abandon earlier achievements, would look like a sure thing. The troubles start when we consider certain cases of theory replacement.

15.4.2 Theory Replacement

The story of the shift from a geocentric (earth-centered) to a heliocentric (sun-centered) model of the solar system is probably familiar enough for us to skip some details. After the Ptolemaic geocentric model had reigned for centuries (from about 200 to about 1550 C.E.), it became increasingly clear that an accumulation of modifications might not be sufficient to keep it empirically adequate. It was Nicolaus Copernicus in particular who complained about the baroque structure of the model, which by then included not only epicycles, but epicycles upon epicycles, and other monstrosities. The real scandal though was that the increasingly complicated model proved increasingly useless, especially for purposes of navigation, which had taken on greater importance with the discovery of the New World. A better model was needed.

Probably inspired by new developments in Islamic astronomy, as recent research suggests, Copernicus had the brilliant idea to radically change one of the basic, ancient modeling assumptions. Instead of leaving the earth at the center of the universe, he produced a model in which the sun took center stage. An entirely new theory was born. Officially, Copernicus assured his readers that he only proposed a model that simplified the computations of celestial positions, and consequently navigation. He, or at least his student Andreas Osiander, who saw the book through its printing after Copernicus died, took an instrumentalist attitude toward the model – again, at least officially. Be that as it may, the model slowly gained the command of the entire astronomical community and eventually became, after further, important modifications (i.e., Kepler's elliptical orbits), accepted as the correct model of the solar system.

This shift from geo- to heliocentrism is, intuitively, a clear example of theory (or model) replacement, not merely modification. It is a bit tricky, though, to pinpoint exactly when a series of modifications to an existing theory result in a new theory. Why, one might ask, was the introduction of epicycles upon epicycles a mere modification of the basic Ptolemaic model, while switching the positions of Earth and Sun a full-blown replacement? One might say that geocentrism was a pillar of the Ptolemaic model, so that once it was given up, a new model emerged.

However, according to most historians of science, the third assumption of circular orbits was equally central, and yet we usually don't think of Kepler's elliptical model as replacing the Copernican model, but instead we consider it a modification. However, a clear-cut distinction between modifications and replacements is neither possible nor important for understanding the changing relations between theories/models and the world. Suffice it to say that along the spectrum from modification to replacement, there seem to be clear examples of both.

A purely progressive picture of the history of science runs into trouble with theory replacement. According to the progressive picture, science just starts from humble beginnings (simple models) and develops through modifications of those models in response to evidence. What doesn't happen, on this naïve view, is that we ever abandon earlier achievements. However, the phenomenon of theory replacement shows that this can't be right. We don't think the geocentric model was "on to something," but rather, we think that it got things pretty badly wrong. There are many other examples that illustrate the phenomenon of abandoning old theories, instead of just modifying them: In chemistry, the phlogiston theory of combustion was replaced by the oxygen theory; in biology, the "theory" of de novo creation of the different species was replaced by the theory of evolution; in geology, the theory that earthquakes are solely the result of volcanic activity was replaced by plate tectonics; in economics, the labor theory of commercial value was replaced by the theory of supply-and-demand; and so on. The history of science is littered with dead theories. The simple progressive account of science often is just plain wrong.

15.5 What Kind of Progress?

However, maybe we can modify the simple progressive model and salvage the idea of linear progress. Although many of our past theories turned out to be false, the new theories that replaced them are better, and thus progressive, in that they are *closer to the truth*. The Copernicus-Kepler model clearly delivered the truth about the structure of our solar system. In this case, the new model triumphed over the old one because the old one could be shown to be false. Similarly for plate tectonics, the micromotion theory of heat, and quantum mechanics: All of them get us closer to the truth about the world, even if not all the way there. That obviously is sufficient for scientific progress.

This line of reasoning embodies the realist attitude that we discussed in Chapter 13. Realism is the view that our best theories are at least approximately true. The view faces a host of problems, of which the underdetermination of theory by data and the pessimistic induction are just the two most serious. Those problems call into question any quick inference from the empirical adequacy of a theory to its truth. This suggests that progress should perhaps not be tied to truth. Maybe the history of science is simply a history of increasing empirical adequacy? For many scientists, that might very well be sufficient. After all, if all we want from our theories is prediction and control, empirical adequacy is all that is needed. On this pragmatic view, progress simply consists in the development of theories and/or models that are increasingly empirically adequate. Scientific optimism is justified, even if we have to abandon our longing for deep truths about the universe, including unobservables. As long as we can make certain that newer theories fit the observable world better than the old ones in ways that provide prediction and control, we have progress. This is where Thomas Kuhn throws in a monkey wrench: What if the world to which we fit our theories is itself partly determined by those very theories? And if that is true, i.e., if there is no theory-independent world, then what sense does it make to say that newer theories *fit the world better* than the old ones?

15.5.1 Scientific Revolutions

Thomas Kuhn, a physicist turned historian of science, was impressed by the occasional occurrence of theory replacements that brought in their wake a *revolution* of our entire scientific outlook. He thought that such revolutions can change the way in which the world is conceptualized, and consequently, the way in which it is observed. He also thought that the conceptualized world is the only world that matters for science, as we never deal with a nonconceptualized world. Thus, revolutions change the world to which our theories must fit. This has the dreaded consequence that we can't define increasing empirical adequacy – the world to which our theories fit is not a stable bedrock that we could use to measure progress in empirical adequacy. Let's start by looking at why one might think that some of the theory replacements mentioned above are indeed revolutions.

Consider the shift from geo- to heliocentrism. First, Earth no longer has a special place in the universe; it becomes just one among many other heavenly bodies. Second, in the course of improving the Copernican model, and due to other developments, the ancient categorical distinction between what happens on Earth (lots of change) and what happens in the region starting with the moon (eternal circular motion of unchanging celestial objects) begins to break down. Third, the motions we observe in the universe are in part constituted by the motion of the earth, which means that fourth, we become, for the first time, aware of the fact that what we observe is not always a simple record of what is happening independently of us.

Or consider the publication of Charles Darwin's *The Origin of Species* in 1859, which introduced a unified theory of evolution. Instead of the then fairly widespread assumption that all the different species were created on a case-by-case basis (a process called *special creation*), Darwin proposed that all contemporary species are descendants from a common ancestor through a process involving heritable variation and natural selection over long periods of time. This was truly revolutionary. First, it was thoroughly naturalistic – the explanation for the variety of species we observe does not involve any supernatural causes. Second, it was nonteleological: There was no final goal or apex of creation. Even humans were not special but simply evolved *from* earlier origins, and were *not* hypothesized to evolve *toward* some end state. And third, selection of favorable traits did not presuppose an intelligence of any kind (as it does in breeding); scarcity of resources and heritable variations are enough for selection to take place.

15.5.2 Kuhn's Cyclical Model

In 1962, Thomas Kuhn published a revolutionary, small book, *The Structure of Scientific Revolutions*. In this book, Kuhn dealt a blow to the traditional view of the history of science from which it has not recovered. Roughly, he argued that there is no linear progress discernible in the history of science, for the simple reason that what counts as scientific truth changes radically during the revolutions that characterize the history of science. These revolutions include the *Copernican Revolution* (1543), the *Chemical Revolution* (1789), the *Special and General Theory of Relativity* (1905 and 1915, respectively), and *Quantum Mechanics* (early twentieth century). Kuhn replaced a linear picture of the development of science with a cyclical one, postulating the stages shown in Figure 15.1:

The infancy of any science is called its *preparadigm* period. During this period, researchers are trying hard to find the right approach to loosely connected phenomena. At first, it's entirely unclear which phenomena should

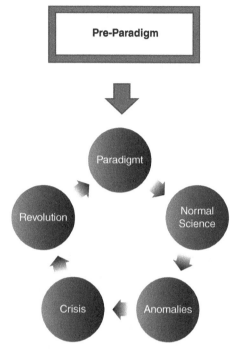

Figure 15.1 Kuhn's model. [ME note: reproduce this in mono].

be grouped together. It's equally unclear what could count as evidence for or against a theory about the phenomena, as well as what explanatory strategies are fair game: Should the phenomena be explained in terms of action-at-a-distance or in terms of mechanical interactions? Are invisible substances responsible for the observed regularities, or are there strange forces?

This chaotic state of affairs is only resolved once a particular approach proves itself to be much better than all of its competitors. It becomes the *paradigm* for how to do research concerning a now clearly delineated class of phenomena. A good example is Classical Mechanics, which, after the publication of Newton's *Mathematical Principles of Natural Philosophy* in 1687, became the model for how to study the physics of motion. A paradigm consequently ushers in a long period of *normal science*, during which it is further articulated: More powerful mathematical tools are developed in the course of applying its principles to new cases, and the values of important constants are determined to ever higher degrees of precision. Such periods of articulating a paradigmatic theory can be found, not only in physics, but also in chemistry, biology, geology, etc. During such tranquil periods, scientists are busy with "normal science," tweaking and patching paradigmatic theories to increase their scope and applicability.

Inevitably, the good times come to an end. The paradigm faces an increasing number of *anomalies*. These are phenomena that the theories within the paradigm simply cannot account for, although everybody agrees that they should be able to do so. Tweaking theories just doesn't do the trick anymore. The whole paradigm – the role model for how to do science concerning a well-delineated set of phenomena – is in *crisis*. Old questions, the answers to which had largely been taken for granted, rear their ugly heads again: "Maybe we should allow instantaneous action-at-a-distance? Maybe the past *can* be influenced by the future?" How theories are tested is starting to become itself contested, and scientists divide into different camps, each with its own ideas of how to proceed.

These camps become increasingly competitive. They start their own exclusive meetings, maybe even found their own journals. Especially from the outside, the competition starts to look like a regular old brawl. We have entered the period of *revolution*. Scientists from different camps have a difficult time arguing with each other in a rational manner, as they no longer even agree on what the relevant evidence is, what sort of approaches are legitimate from a methodological perspective, and what sorts of metaphysical commitments are acceptable. Memories of the preparadigm period emerge, and with them the sense that only the establishment of a new paradigm will settle things. But which camp will provide that new paradigm?

Kuhn's answer to this last question has upset many in the scientific community, as well as many philosophers of science. In a nutshell, his answer is this: The winner of a revolution is not just determined by the evidence, the fruitfulness and plausibility of the theories proposed, or by standard means of theory testing. The winner is, rather, to a large extent determined by social and political forces. Scientific rationality seems to play a minor role. Scientific revolutions are, in this sense, nonscientific.

15.5.3 The Rationality of Revolutions

One important reason for Kuhn's view is his reliance on a particular model of scientific rationality that breaks down during revolutions. According to this model, rationality in theory choice requires an agreement on the relevant evidence and on what constitutes a rigorous empirical test of a theory. But these factors are disputed during a revolution, which means that the necessary ingredients for scientific rationality are under dispute. Thus, when revolutions eventually yield a winner, the driving factors must be unrelated to scientific rationality. They must be factors that are not rational in the scientific sense, such as social and economic forces as well as political allegiances (Kuhn explicitly compares scientific with political revolutions).

There are several ways to react to the Kuhnian picture of what happens during revolutions. One possible reaction is to declare the search for truth as a misguided ideal for science. If revolutions are settled by nonscientific means, maybe philosophers of science should generalize this insight and study social, political, and economic power structures as the driving forces behind all scientific developments. Much of the so-called *science studies* and *constructivism* about science can be seen to result from this interpretation of Kuhn's provocative picture. At their most radical, these trends have sometimes resulted in claims denying the very possibility of objective scientific insights into the structure of the empirically accessible world. Scientific theories, on such a view, are purely social constructs and best understood in terms of underlying power structures, and thus have no better claim to objectivity than other forms of social narratives.

An alternative reaction to Kuhn's analysis of scientific revolutions is to question his choice of the model of scientific rationality. True, an essential part of scientific rationality is to propose bold and testable theories and then let them face the tribunal of experiments. But we have already seen repeatedly that this is not sufficient. Given, for example, the underdetermination of theories by data, nonempirical considerations will have to be taken into account in order to make theory choice possible. Scientists might invoke considerations concerning accuracy, consistency, breadth, simplicity, fruitfulness, conservatism, or aesthetics. Doing so need not necessarily be irrational. In addition, scientists might bring to bear conceptual considerations more familiar to philosophers.

For example, faced with a choice between the phlogiston theory of combustion and the oxygen theory, one might point to serious conceptual difficulties with the former. Remember that Lavoisier had shown that mercury, when burned, actually gains weight. Defenders of the phlogiston theory accounted for this by proposing that phlogiston has negative weight, so that an object from which it is released becomes heavier. Even if we assume that this modified phlogiston theory can be formulated in an empirically adequate way, one could still argue that the phlogiston theory is conceptually defective. It is part of our concept of a physical substance that it has mass, which translates into weight in an appropriate gravitational field. Thus, a substance with negative weight should have negative mass, and it seems that the concept of negative mass is utterly confused.

We are not claiming that this brief example shows how all stand-offs between empirically equivalent theories can be resolved. It is simply intended as an illustration of how conceptual considerations might play a role in the development of science that is overlooked by the somewhat naïve model of scientific rationality sketched above. The idea is that if empirical evidence alone cannot decide between rival theories, then comparing them from a conceptual perspective migh be fruitful. One shouldn't be too optimistic, though. Especially revolutionary changes often bring in their wake conceptual upheavals. In the general theory of relativity, spacetime is curved. In quantum mechanics, nonlocal causation might be required. Plate tectonics makes entire continents move. Philosopher Larry Laudan suggested that we can take account of such conceptual upheavals in our assessment of progress by incorporating them into our determination of the problem-solving effectiveness of what he calls a research tradition. To do this, we need to estimate the number and importance of problems solvable within a tradition and subtract from it the number and importance of anomalies and conceptual problems produced. As can easily be imagined, coming up with such estimates is not going to be a precise enterprise, but a more detailed exploration of this proposal is beyond the scope of our book.

15.6 From Theories to Research Programmes

Another possible reaction to Kuhn's challenge that has attracted a lot of attention is Imre Lakatos' *methodology of research programmes*. It starts with the observation that research programmes (his term for paradigms) are always embedded in a context that sees many different such programmes being pursued at the same time. The problem of the rationality of revolutions now becomes the problem of the rationality of switching from one research programme to a competing one. Lakatos distinguishes between progressive and degenerating programmes. A programme is progressive to the extent to which modifications of its theories or models in light of new evidence leads to the prediction of *novel* facts and empirical confirmation of those predictions. If only novel facts are predicted, the change is called "theoretically progressive," and if some of those predictions are borne out by the evidence, it is also "empirically progressive." A research programme is called "degenerative" if it is neither theoretically nor empirically progressive. Lakatos claims it is rational to switch from degenerative to progressive research programmes.

Much of the support for this claim is derived from the history of science. As Lakatos points out, many important programmes that intuitively were worth pursuing were *born refuted*. For example, Bohr's early theory of the emission of light was based on inconsistent models. On the one hand, he adopted Ernest Rutherford's model of atoms as miniature solar systems with electrons playing the role of the planets that orbit the nucleus. On the other hand, the then accepted Maxwell-Lorenz theory of electromagnetism entails that an atom so modeled should collapse. If Popper were right about falsification being sufficient for rejecting a theory, this inconsistency should have led to the immediate abandonment of Bohr's theory. But it didn't – and rightly so, says Lakatos. He distinguishes between the *hard core* of a research programme and its *protective belt*, which consists of various auxiliary hypotheses. If a theory within the programme ends up conflicting with facts (which, for Lakatos, are the results of interpretations of evidence in light of further theories), we first attempt to reinstate consistency by modifying the hypotheses of the protective belt, which include interpretive theories. And if this can be done in an either theoretically or even empirically progressive way, we continue with our allegiance to the hard core of the programme. This is what happened in Bohr's case. Ignoring the inconsistencies, he started to construct a model of the hydrogen atom, which included a proton-nucleus and one electron in a circular orbit. Next, he explored elliptical orbits, then tried to model more complicated atoms and their interactions with electromagnetic fields until … until the fixes needed to make the theory at least somewhat empirically adequate became increasingly less progressive. Eventually, Bohr's programme was replaced by wave mechanics, which had all the trappings of a progressive programme.

There are some parallels between Lakatos' and Kuhn's approach. Importantly, both agree that Popper's model of science, which requires the immediate abandoning of falsified theories, is itself falsified by the history of science. Kuhn thinks that anomalies need to first endure for some time by resisting all attempts to fix inconsistencies between predictions and facts before a crisis ensues, which eventually leads to a scientific revolution. Similarly, Lakatos observes continuing changes in the protective belt in order to protect the hard core of a research programme, and only when such changes become degenerative should the programme be abandoned. However, there are also deep disagreements between the two. While Kuhn, at his most radical, seems to think that the forces deciding a revolution are largely social and political, Lakatos aligns with the spirit of Popper's *critical rationalism*: The growth of scientific knowledge is the product of continuous critical activity taking into account empirical evidence, the strength of competing theories, and the innovativeness of theoretical modifications. If it is in the nature of science to aim at increasingly better theories, i.e., theories that contribute to the growth of knowledge, the switch from degenerative to progressive research programmes looks indeed rational.

Related to this view is a second disagreement between Kuhn and Lakatos. Kuhn tends to conceptualize the history of science as a *sequence* of revolutions, normal science, crises, new revolutions, and a new normal science. In this picture, new conceptual and theoretical developments only happen *after* a dominant paradigm experiences a crisis. In contrast, Lakatos holds that new developments happen all the time, even while a dominant research programme is fully flourishing, with no signs of trouble on the horizon. Thus, it can happen that a new, fledgling programme flies under the radar, only to become a prominent competitor once the dominant player is in trouble. Moreover, trouble can even arise from the existence of such a fledgling programme: News of its success in better accommodating some evidence that is difficult for the dominant programme to account for makes the rounds, and soon the fledgling competitor puts additional pressure on the dominant programme. Not only inconsistencies with facts spell trouble for a programme, but also the existence of competitors that are more successful in handling those facts. Neither Popper nor Kuhn make room for these additional pressures. But somebody else does: By a radical extension of Lakatos' emphasis on competition, Paul Feyerabend ends up agreeing with Kuhn that there are no eternal rationality standards that determine the outcome of revolutions, small or large.

15.7 Methodological Anarchism

In his (in)famous book from 1975, *Against Method*, Feyerabend embarks on a far-reaching attempt to demolish the hegemony of allegedly unchanging rationality standards used for evaluating scientific progress and for demarcating science from pseudoscience. His principal claim is that if scientists were to adhere by the methodological rules dreamed up by philosophers, the science we know would cease to exist, or would never have developed in the first place. He rests his case largely on a detailed historical study of Galileo's defense of Copernican heliocentrism, including his employment of the telescope, which Feyerabend claims to have been dubious at best in the historical context. If the methodological norms defended by Popper and others had been adhered to, Galileo would have never prevailed. Rather, so Feyerabend contends, it looks as if Galileo's and Copernicus's adversaries used methodologically sound arguments, while Galileo resorted to parlor tricks, political coat-tail riding, and rhetorical ruses. Thus, given that Galileo turned out to be right, and given that adherence to the rules of something like critical rationalism would have killed off his theories, we should throw out those rules. Anything goes!

The interesting question is how far methodological pluralism can in principle be extended. We have already seen that Lakatos endorses constant competition among programmes that often use different methodologies. In his most radical moments, Feyerabend seems to want to go further. Why stick with the rules of science? Why not use astrology, shamanism, or treat ancient myths as sources for knowledge about the world? He marshals two considerations in favor of this attitude. First, looking at the history of human attempts to formulate worldviews, he observes many different traditions engaged in such worldview making efforts and argues that privileging one tradition, Western science, appears to be arbitrary. In particular, even celebrated episodes in the history of science, such as Galileo's defense of Copernicanism, are less constrained by strict canons of rationality than is usually thought. Second, and a little more reasonable, he thinks that the best way to test standard scientific methodology is to allow for research that doesn't adhere to those rules.

It seems to us that Feyerabend accuses philosophers of Western science of promoting a Wiggish rewriting of the history of attempts to understand the world: First, they abstract general methodological principles from the current practice of science (Popper's falsificationism is his favorite target), and second, they then turn around and select as worthwhile only those traditions that seem to embody those principles. But, he emphasizes, this is wrong on two counts: The history of scientific success is often incompatible with those principles, and the only real way to test the efficacy of those principles is to have research carried out in accordance with them compete with research that violates them. In a way, he remains a falsificationist in the style of Lakatos, not about research programmes, but about methodologies and even standards of rationality.

Historians of science have rightly accused Feyerabend of cherry picking from the relevant historical texts in order to make his case. For example, while Feyerabend argues that Galileo's use of the telescope for astronomical observations constituted an illegitimate application of an instrument that had only been validated on Earth, he ignores the extended argument for the use of the telescope that Galileo provides in his discussion of the moons of Jupiter. Perhaps more damning for the *anything goes* slogan is the fact that during the history of humankind, many alternatives to historically dominant practices have been tried, and many of those have resulted in failure. To this date, no plausible studies exist that show any efficacy of so-called homeopathic treatments of illnesses. Astrological predictions of a person's character on the basis of her birth chart are notoriously wrong. Young-earth creationism is both explanatorily and predictively barren. The examples of failed traditions could be multiplied.

In fairness to Feyerabend we need to point out that in later writings, he walked back some of his more radical claims from *Against Method*, saying that he simply liked to be provocative. What he really wanted to promote was a methodological pluralism within science itself. The fear is that by forcing science into a methodological straightjacket, we run the risk of stagnation. There is, of course, some truth in this. Looking at science funding, the rules for submitting grant applications could easily lead one to believe that only science done in a very particular and regimented way is good science. The danger of missing some important new approach is all too real. Perhaps we should read Feyerabend as issuing a somewhat exaggerated, and certainly provocative, warning that needs to be heeded.

Apart from his agreement with Kuhn on the importance of recognizing social and political forces that are operative in science, Feyerabend also argued for another infamous doctrine attributed to Kuhn's work, viz., the *incommensurability* of scientific terms from different paradigms. Such incommensurability of meaning entails that we cannot say that a theory from one paradigm is closer to the truth than a theory from another paradigm. This happens because the incommensurability of terms results in an incomparability of the worlds the theories address.

15.8 Incommensurability

The thesis of the incommensurability of theories challenges the views of both the veritistic and the epistemic view of scientific progress that newer theories are closer to the truth than older ones, or are at least empirically more adequate. This challenge is grounded in the claim that such theories don't talk about the same things. Even if they use the same terms, such as "mass," what those terms pick out changes during scientific revolutions. It's as if two people argue over which the best football team is – the Denver

Broncos or Manchester United. If they were serious in their argument, it would have to be the case that they don't recognize that when they use the term "football," they mean different things – one guy is talking about American football, while the other one is talking about what the Americans, but not the British, call "soccer." Given the difference in the meanings of the very same term, they can both be right, and of course they can both be wrong. But there is no point trying to compare what they are saying, in the sense of trying to figure out who is closer to the truth. They use the term "football" with incommensurable meanings, which makes their theories about the best football team incommensurable as well.

Kuhn argues that a similar situation can, and often does, arise for scientific terms, such as "mass," "atom," and so on. Consider the term "mass," as it is used in classical mechanics. It stands for something like "quantity of matter" and figures in equations such as $F = ma$, or $p = mv$. The value of m doesn't change as a function of an object's motion. In the special theory of relativity, it does change. It increases as the velocity of an object approaches the speed of light. The question is: Does "mass" in classical mechanics stand for the same thing as it does in the special theory of relativity?

In this case, one might think that the answer is a simple yes. It's just that we have found out something about mass that Newton and other proponents of classical mechanics weren't aware of, viz. that its value can change as a function of velocity. However, the story gets much more complicated when we are dealing with theoretical terms. As we have seen previously, theoretical terms are tricky. We cannot introduce them with a definite meaning by pointing at their referents, in the way we introduce names into our language. Theoretical terms refer to entities that are not directly observable. According to one popular view, we introduce them into a theory by specifying a set of properties that the referents must have. For example, "atom" refers, in ancient Greek theory, to those smallest constituents of matter that can no longer be divided and combinations of which give rise to all material substances. Thus, when we talk about atoms in ancient Athens, we are talking about that which is indivisible. There is a conceptual connection between atoms and indivisibility. From that perspective, postulating a theory that allows for the "splitting of atoms" is simply nonsense.

Now imagine transporting a scientist who works at CERN, the big particle accelerator in Switzerland, back in time to ancient Athens. After her arrival, she goes to the agora (the main plaza) in search for some food and drink. Her modern outfit causes consternation among the Athenians, but one of them finds the courage to sit down next to her and engage her in a conversation. After a while, he bursts out laughing: "You did what? You split an atom?!" She nods her head, insisting that splitting atoms is daily routine at CERN. Seemingly, the two disagree on whether or not atoms can be split, just as the two guys from earlier disagreed over the best football team.

The way to end such disagreements is to point out to the parties involved that they are talking past each other. "Football" means one thing in American minds, quite a different thing in British ones. And "atom" means one thing in ancient Athenian minds, and quite a different thing in those of CERN scientists. The meanings are incommensurable, and so are the theories.

In slightly more technical terms, the argument for incommensurability goes something like this. The meaning of the theoretical terms of some theory T is fixed by their *inferential relations* to all other terms in T. For example, if a Greek scientist identifies something as an atom, it can be inferred that it is indivisible. Therefore, the meaning of "atom" entails indivisibility. It follows that if any of those inferential relations change, so do the meanings of the affected theoretical terms. If we add a form of holism, according to which *all* terms in a theory are inferentially connected to all other terms of T, then any change in T changes the meanings of all terms of T. If this is the case, all theory changes results in incommensurability.

One proposal for resolving this conundrum is to abandon holism and restrict the relevance of inferential connections to theoretical terms, but not extend them to observational terms, thus rescuing their stable meanings from changes in theory. This proposal to anchor the meaning of some terms in observation is, however, under attack from another idea found in Kuhn: The idea that different paradigms generate different observable worlds.

15.8.1 New Worlds?

An intuitively plausible way to avoid a complete dependence of the meanings of scientific terms on the theory in which they are deployed is to insist that some of those terms have meanings grounded in observation. Earlier, we called such terms *observational terms*. The idea was that some terms refer to observable things; the term "pendulum," for example, simply refers to observable pendula, such as inside a grandfather clock. While we have seen that it can be tricky to draw a principled distinction between observable and nonobservable entities, and thus between observational and theoretical terms, the mere existence of clear examples of observables provides an anchor for the meaning of at least some scientific terms. Moreover, this anchor promises to solve the problem of incommensurability. Even if the meaning of theoretical terms is determined by their inferential connections to other terms in the theory, this is not the case for the meaning of observational terms, which is determined (in part) by the observable entities to which they refer. Thus, theoretical changes do not change the meaning of observational terms. Furthermore, this stability seems to span revolutions and thus bridge the gap between paradigms – after all, the observational world we try to account for in our empirical theories is unaffected by our theoretical beliefs about it.

Nice try, says Kuhn, but not so fast. First, remember the consequences of the theory-ladenness of observation. On both the strong and the weak version of this thesis, what we observe depends in part on the background theories we already accept. Sometimes, those theories do not amount to much more than low-level beliefs about the context in which one makes an observation: Seeing identical cuts to a person's skin counts as observing a stab wound after a bar fight and as a surgical incision in the operating theater. In other cases, more elaborate theories come into play. The expert's observation of protons in a Wilson Cloud Chamber provides another example (see the mysterious image in Chapter 4). Only in light of a well-articulated background theory concerning the interactions among elementary particles do the visual impressions of various lines count as observing the signatures of particles.

Kuhn takes the impact of theory-ladenness a step further and emphasizes its effects after changes in paradigms, i.e., after scientific revolutions. Not only do such revolutions result in radically different beliefs about the world, they bring about radical changes in how the world is conceptualized at the most basic level. To use Kuhn's famous example, when Aristotle looked at a bob swinging on a string, he saw an object engaged in constrained "free fall," but after Galileo we see a pendulum exhibiting circular motion. In this case, Aristotle's binary distinction between natural and forced motion was replaced by a larger class of different motions – linear motion, accelerated motion, circular motion, uniform motion, and various combinations thereof. Thus, while the motions observable by Aristotle fell into two jointly exhaustive categories, the motions we can observe are of much greater variety. Kuhn concludes on the basis of considerations such as this that the world – the *observable* world – of Aristotelian physics is a different kind of world than the one that came into existence in the wake of the scientific revolution of the sixteenth and seventeenth centuries.

If this is right, we face a radical consequence: *What it means to be a true theory changes after a scientific revolution.* To see this, recall our brief discussion of truth in Chapter 1. The simplest account of truth is embodied in the *correspondence theory*, according to which an empirical theory is true just in case it corresponds to the facts of the world, including, of course, the observable facts. However, if, as Kuhn argues, the observable facts change after a scientific revolution, then the "things" a theory has to correspond to in order to be true are different! This has two consequences. First, incommensurability seems to be truly inescapable because even observational terms are affected by paradigm changes. Thus, secondly, new paradigms result in new worlds of observation. There is, in short, *no accessible stable world* behind those different observable worlds that are the result of how paradigms determine what we see.

These are truly radical consequences of Kuhn's picture. As you might expect, this picture doesn't sit easily with many in the scientific community. In particular, it would seem that the claim that the observable world itself is subject to changes during scientific revolutions is simply outrageous. After all, there is a sense in which Aristotle would see a pendulum, were he to be taken from the past and looked at a contemporary grandfather clock. Granted, he wouldn't see it *as* a pendulum, but why would that be important?

This line of thought has been discussed, not so much in the philosophy of science literature, but in the philosophy of mind and language, under the topic of *nonconceptual content*. While we obviously cannot explore this discussion in depth, here's the main idea. Most of our mental representations have conceptual content. When I look out of the window, I represent the scene I am seeing as various people walking across a lawn. Described this way, my mental representation of the scene outside my window is thoroughly conceptualized. I see *people*, who are *walking across* a *lawn*. Observing a person walking across a lawn is the result of, among other things, applying the concepts of "person," (as opposed to "dog"), "walking" (as opposed to "running"), and "lawn" (as opposed to "prairie") to visual impressions. Thus, the content of my representation of the scene outside my window is *conceptual content*. The question now is whether or not nonconceptual contents also exist. If they did, maybe they would be candidates for representing a world that is independent from background theories, which usually are the source of the conceptual part of the content of our representations. In other words, maybe the observable world that remains stable throughout all changes in the worlds of science is the world as it is represented nonconceptually.

The possible existence of nonconceptual content is highly controversial. One frequently purported example is the content of our representations of color. It is well known that humans can visually discriminate millions of different hues. Obviously, by far the largest part of those distinctions have no corresponding distinctions in our repertoire of color concepts. We simply can distinguish many, many fewer colors conceptually than we can perceptually. This has been taken to suggest that the difference in content between looking at two different hues that cannot be marked with a conceptual distinction must be nonconceptual. It is unclear though whether or not this simply shows that we have far fewer color *terms* than representations of colors. In other words, could it be the case that even visual color discriminations are conceptually influenced, despite the fact that we lack the corresponding words? Obviously, this is a question that falls squarely within the domain of cognitive science, and we won't speculate on it here any further.

At any rate, the possible existence of nonconceptual content is of dubious utility to the opponent of Kuhn's suggestion that each scientific paradigm comes with its own observable worlds in a package deal. Science is rarely, if ever, concerned with the world as it is represented by mental states with nonconceptual content. On the contrary, the observable objects for any ordinary empirical science are objects *seen as* instances of certain concepts: levers, atoms, organisms, chemical bonds, market trends, and so on. If Kuhn is correct in his claim that the concepts used to categorize the observable world into discrete entities are subject to radical remakings in the wake of scientific revolutions, it is very difficult to escape the consequences of the incommensurability between theories. In particular, the arguments against traditional forms of scientific realism, many of which turn on the nonprogressive nature of science, receive powerful support from this view. Moreover, if incommensurability is as pervasive as both Kuhn and Feyerabend claim, what does this entail for the notion of scientific progress, something dear to the heart of realists?

15.9 Structural Realism and Progress

Maybe the realist can retreat to a weaker position inspired by the observation that sometimes at least, two theories from different paradigms are related to each other through the existence of limiting procedures, as outlined earlier. There, we illustrated the notion of a *reduction in the limit of some quantity* by using as our examples classical mechanics and relativity theory. These two theories are of course divided by one of the most far-reaching scientific revolutions in the history of science. And yet, it seems that certain *mathematical structures* exhibit a continuity that spans the revolutionary divide.

The sort of realism inspired by this observation is called *structural realism*. Recall from Chapter 12 what it says: Roughly, while theories often change radically in what they say about the *inhabitants* of certain (mathematical) structures, those structures themselves are relatively stable. For example, classical mechanics entails that the inhabitant m of the structure defined by formula (1) earlier in this chapter, the property we call "mass," is such that its value is independent from motion. Relativity theory on the other hand entails that the inhabitant m can change in value with motion. The structural realist concludes that what we should be *realists about* is *the mathematical structure* common to both theories, where in this case, the structure postulated by classical mechanics turns out to be a special case of the structure postulated by relativity theory. But we should be *agnostic about* the postulated *nature of the inhabitants* of those structures.

This view is often characterized as retaining the best of both worlds – realism and antirealism. As a realist about the structures, one claims to be able to account both for the predictive success of our best theories and for continuity through theory change, even if that change deserves to be characterized as revolutionary. As an antirealist about the nature of the inhabitants of those structure, one is not committed to the truth of what theories tell us about the furniture of the universe. Thus, the force of the pessimistic induction is blunted, and along with it, perhaps also the worries about Kuhnian incommensurabilities. Of course, whether structural realism can successfully combine the best of both worlds will depend on finding sufficiently many and sufficiently stable structural continuities in the history of our empirical sciences. This is a question for much further research.

Returning to our fourfold distinction among different scientific goals that can be used as a yardstick for measuring progress, we have to ask which of those goals, if any, is most compatible with structural realism. First, the *noetic* goal, which measures progress in terms of increases in understanding, appears to be well served by discovering the correct structures. As pointed out earlier, there is a strong tradition in science of seeing our understanding being furthered by the grasp of structural relations. Maxwell's flow analogy is a prominent example of such a view, for the explanatory work is done solely by features of the structure he postulates (a system of sources, sink, and pipes), while the nature of the inhabitants is irrelevant, and thus can be entirely fictional. Second, the *ability* goal, which emphasizes the problem-solving capacity of research traditions, requires that it be possible to formulate the problems to be solved in terms of questions about structures. This is perhaps possible in mathematical physics, but not obviously the case in most of the other empirical sciences, such as biology or psychology. Similarly, it is clear that the *veritistic* and the *epistemic* goals need to be redefined. Rather than getting closer to the truth about the nature of the inhabitants of structures, progressive science only gets closer to the truth about the structures themselves, which are simply collections of relations.

If any of those last three goals are the appropriate ones for scientific activity, structural realism becomes less plausible as an anchor for scientific progress. For example, what exactly are the structures in biology about which it would make sense to be realist? Whatever they are, they are most likely not mathematical in nature (the Hardy-Weinberg equation of population genetics, the Lotka-Volterra equation of predator–prey dynamics, and other quantitative models notwithstanding). And the same is presumably true about many other sciences, such as psychology. These kinds of disciplines seem to be focused on the nature of the inhabitants. For example, is the mind really a system of cognitive functions that mediate between sensory input and behavioral output? Or is it rather a high-level description of lower-level, neuronal interactions? What is the nature of neurons? Should it turn out to be the case that both (i) appropriate structures are absent from disciplines like sociology, psychology, and biology and (ii) the history of those disciplines exhibits revolutionary changes, Kuhn's problems concerning incommensurability could not be solved by structural realism for those disciplines. It is, of course, an open question whether or not all empirical sciences can be transformed into theories in which (mathematical) structures play the central role. It is another question altogether whether such a transformation would even be desirable. And finally, it is an open question whether the life and social sciences undergo Kuhnian revolutions or engage in continuous theory revision.

15.10 Conclusion

As you can see, there are many unresolved questions concerning both the historical development and the future of science. Is there any theoretical stability across scientific revolutions? In what sense is the world independent from the sciences that seek to describe and explain it? Could this world simply be a world of mathematical structures? And as we briefly considered in Chapter 14, what does the future of science hold with respect to interdisciplinary research? It is clear that these questions, and others like them, cannot be answered by empirical methods alone. To be sure, the question of the theory-ladenness of observation, for example, will incorporate results from an empirical science, viz., cognitive psychology. In trying to answer some of the other questions, we need support from linguistics, mathematics, history, and philosophy. Empirical science itself is deeply embedded in a system of intellectual endeavors not all of which are themselves fully empirical. All of our empirical, scientific disciplines are spin-offs from what used to be called natural philosophy. Perhaps we should remember this origin and realize that the sciences are less independent from the entire gamut of human intellectual engagement with the world than is sometimes claimed. What this means for science in the coming decades as diverse disciplines seek to collaborate is that serious consideration of philosophical foundations will be crucial if we hope to solve the complex theoretical and practical problems of the twenty-first century.

Annotated Bibliography

Paul Feyerabend, 1975, *Against Method*. Verso, Brooklyn, NY.
    The fourth edition from 1992 contains an introductory essay by Ian Hacking, Feyerabend's last letter before he died in 1999 (a letter to the reader), Feyerabend's prefaces to the first three editions he saw through the print, and of course the famous "Analytical Index, Being a Sketch of the Main Argument." In its impact on a wide audience, this book rivals, and perhaps exceeds, that of Kuhn's *Structure*. It is written in a somewhat idiosyncratic, but often highly entertaining style, and it is intentionally very provocative.

Philip Kitcher, 1993, *The Advancement of Science. Science without Legend, Objectivity without Illusions*. Oxford University Press, Oxford, UK.
    The author introduces a model for scientific progress based on an analysis of microchanges. It takes into account cognitive, economic, and psychological aspects of science, for an adequate treatment of which we didn't have the space. The book starts with a detailed analysis of the success of Darwin's new theory of evolution, and it ends with an important discussion of the organization of cognitive labor within the sciences.

Thomas Kuhn, 1962, *The Structure of Scientific Revolutions*. The University of Chicago Press, Chicago, IL.
    The 50th Anniversary Edition from 2012 contains a very helpful introductory essay by Ian Hacking, which places Kuhn's work into its historical context and also provides the reader with a good impression of how it was received by the academic community. It is very unusual for a book in philosophy to have such an extreme impact on other disciplines, as Kuhn's *Structure* did and still does.

INDEX

This is Philosophy of Science: An Introduction, First Edition. Franz-Peter Griesmaier and
Jeffrey A. Lockwood.
© 2022 John Wiley & Sons, Inc. Published 2022 by John Wiley & Sons, Inc.